WORLD BANK TECHNICAL PAPER NUMBER 246
IRRIGATION AND DRAINAGE SERIES

Modern Water Control in Irrigation

Concepts, Issues, and Applications

Hervé Plusquellec, Charles Burt, and Hans W. Wolter

The World Bank
Washington, D.C.

Technical Papers are published to communicate the results of the Bank's work to the development community with the least possible delay. The typescript of this paper therefore has not been prepared in accordance with the procedures appropriate to formal printed texts, and the World Bank accepts no responsibility for errors. Some sources cited in this paper may be informal documents that are not readily available.

The findings, interpretations, and conclusions expressed in this paper are entirely those of the author(s) and should not be attributed in any manner to the World Bank, to its affiliated organizations, or to members of its Board of Executive Directors or the countries they represent. The World Bank does not guarantee the accuracy of the data included in this publication and accepts no responsibility whatsoever for any consequence of their use. The boundaries, colors, denominations, and other information shown on any map in this volume do not imply on the part of the World Bank Group any judgment on the legal status of any territory or the endorsement or acceptance of such boundaries.

The complete backlist of publications from the World Bank is shown in the annual *Index of Publications*, which contains an alphabetical title list (with full ordering information) and indexes of subjects, authors, and countries and regions. The latest edition is available free of charge from the Distribution Unit, Office of the Publisher, The World Bank, 1818 H Street, N.W., Washington, D.C. 20433, U.S.A., or from Publications, The World Bank, 66, avenue d'Iéna, 75116 Paris, France.

Cover photos: The top photograph was taken by Hervé Plusquellec in the Guilan Project in Iran. The bottom photograph was taken by Hans Wolter in the Imperial Valley Irrigation District in California.

ISSN: 0253-7494

Hervé Plusquellec is the irrigation advisor to the Natural Resources Division of the Agriculture and Natural Resources Department at the World Bank. Charles Burt is director of the Irrigation and Training Research Center at California Polytechnic State University in San Luis Obispo, California. Hans W. Wolter is theme manager for the International Program for Technology Research in Irrigation and Drainage in the Natural Resources Division at the World Bank.

Library of Congress Cataloging-in-Publication Data

Plusquellec, Hervé L., 1935–
Modern water control in irrigation : concepts, issues, and applications / Hervé Plusquellec, Charles Burt, and Hans W. Wolter.
p. cm. — (World Bank technical paper, ISSN 0253-7494 ; 246. Irrigation and drainage series)
Includes bibliographical references (p.).
ISBN 0-8213-2819-0
1. Irrigation engineering—Technological innovations. I. Burt, Charles, 1943– . II. Wolter, Hans W., 1939– . III. Title. IV. Series: World Bank technical paper ; no. 246. V. Series: World Bank technical paper. Irrigation and drainage series.
TC809.P58 1994
627'.52—dc20

94-3821
CIP

IRRIGATION AND DRAINAGE SERIES

Technical Paper Series

No. 40	Plusquellec and Wickham, *Irrigation Design and Management: Experience in Thailand*
No. 51	Shuval and others, *Wastewater Irrigation: Health Effects and Technical Solutions*
No. 64	Hillel, *The Efficient Use of Water in Irrigation: Principles and Practices for Improving Irrigation in Arid and Semiarid Regions*
No. 71	Mahmood, *Reservoir Sedimentation: Impact, Extent, and Mitigation*
No. 94	Le Moigne, Barghouti, and Plusquellec, *Technological and Institutional Innovation in Irrigation*
No. 99	International Commission on Irrigation and Drainage, *Planning the Management, Operation, and Maintenance of Irrigation and Drainage Systems: A Guide for the Preparation of Strategies and Manuals (also in French, 99F)*
No. 110	Dixon, Talbot, and Le Moigne, *Dams and the Environment: Considerations in World Bank Projects*
No. 115	Le Moigne, Plusquellec, and Barghouti, *Dam Safety and the Environment*
No. 120	Plusquellec, *The Gezira Irrigation Scheme in Sudan: Objectives, Design and Performance*
No. 123	Barghouti and Le Moigne, *Irrigation in Sub-Saharan Africa: The Development of Public and Private Systems*
No. 129	Berkoff, *Irrigation Management on the Indo-Gangetic Plain*
No. 157	Critchley, Reij, and Seznec, *Water Harvesting for Plant Production, Volume II: Case Studies and Conclusions for Sub-Saharan Africa*
No. 166	Ahmad and Kutcher, *Irrigation Planning with Environmental Considerations: A Case Study of Pakistan's Indus Basin*
No. 171	Brown and Nooter, *Successful Small-Scale Irrigation in the Sahel*
No. 173	Byrnes, *Water Users Association in World Bank-Assisted Irrigation Projects in Pakistan*
No. 178	Le Moigne, Barghouti, and Garbus, editors, *Developing and Improving Irrigation and Drainage Systems: Selected Papers from World Bank Seminars*
No. 185	Frederiksen, *Drought Planning and Water Efficiency Implications in Water Resources Management*
No. 191	Frederiksen, *Water Resources Institutions: Some Principles and Practices*
No. 201	Van Tuijl, *Improving Water Use in Agriculture: Experiences in the Middle East and North Africa*
No. 205	Xie, Kuffner, and Le Moigne, *Using Water Efficiently: Technological Options*
No. 212	Frederiksen, Berkoff, and Barber, *Water Resources Management in Asia, Volume I: Main Report*
No. 215	Umali, *Irrigation-Induced Salinity: A Growing Problem for Development and the Environment*
No. 233	Frederiksen, Berkoff, and Barber, *Principles and Practices for Dealing with Water Resources Issues*

Contents

FOREWORD

Irrigation is the largest public investment in many countries in the developing world. The World Bank has played an important role in financing irrigation investments, as total World Bank lending for irrigation amounts to 29 billion US dollars in 1991 prices. Today, the demand for agricultural products - food and fiber - is largely met. This success could not have been achieved without the last half-century's investment in irrigation.

Irrigation will continue to play a critical role in our continued ability to feed ourselves. As demand for agricultural produce increases, driven by population growth and rising income, the bulk of increased production will have to come from irrigated lands. Irrigation supplies plants with water which is often the most critical input to production. There are also strong, positive interactions between irrigation and other major sources of agricultural growth: fertilizer, improved seeds, better husbandry, and integrated pest management.

Water is an increasingly scarce resource, requiring careful technical, economic and environmental management. As the demand for water for human and industrial use has escalated, so has the competition for water used for irrigated agriculture. Thus, the challenge for the irrigation sector is: modernization of irrigation systems and practices, drainage and salinity control, greater attention to cost recovery, measures to reduce pollution from agricultural activities, improvements in the level of service and maintenance of existing systems, investment in small-scale irrigation and water harvesting methods.

This paper deals with the first of these challenges: system modernization. New design concepts and modern technologies already exist and have proven their usefulness in many schemes around the world. What is now required is that these concepts and technologies be assessed and utilized on a larger scale. This publication, which is the first of the new World Bank Irrigation and Drainage Series, is intended to stimulate debate among professionals and to increase awareness of the potential of modern technologies for water control and sustainable irrigated agriculture.

Michel Petit
Director
Agriculture and Natural Resources Department

ACKNOWLEDGMENTS

The authors appreciate the encouragement and support received from Michel Petit, Guy Le Moigne and David Steeds during the preparation of this publication. They wish to thank many colleagues for reviewing the draft and offering constructive comments, among them Ashok Subramanian, Shawki Barghouti, John Merriam, Ken Craig, Peter Lee, Jean Goussard, Pierre Rousset, Darell Zimbelmann and many others. Cecilia Gorriz edited the paper and organized the printing.

The support and assistance received is deeply appreciated; the sole responsibility for the content of the paper, however, rests with the authors.

ABSTRACT

The paper is a contribution to the debate on how to transfer modern irrigation concepts and technology to developing countries. In modern schemes, irrigation is provided as a service to users that should be as efficient and convenient as possible. The authors argue that modern irrigation design is a thought process that starts with the definition of a proper operational plan. Scheme layout and equipment selection should be done in light of the operational objectives and the requirements of the farming systems. The paper addresses technological aspects but it is not a design manual. After an initial discussion of objectives and definitions, the paper reviews various elements of irrigation design and presents technological options. Additional design aspects related to maintenance, user participation, cost recovery and sustainability are discussed. Finally, the debate on the new design approach is opened with a presentation of supporting and opposing views.

Part II of the paper presents brief notes on modernized or rehabilitated irrigation schemes throughout the world, describing the characteristics and preliminary results of modernized schemes. An annex provides details of different water control methods. Some photos are attached to illustrate certain aspects of irrigation design.

PART I: CONCEPTS AND JUSTIFICATIONS

1. BACKGROUND AND PURPOSE OF THE STUDY

The Need for a New Approach to Irrigation Design

The authors of this report maintain that there is no choice but to adopt a new approach to the design and engineering of irrigation projects. The needs for improved performance of irrigation projects are pressing, the evidence of the failure of the gated, manually operated systems is overwhelming, and the potential for improvements is undeniable.

Numerous reports published in recent years describe a grim picture of irrigation. Levine and Coward (1986) summarize several points that are repeated in other studies on irrigation projects in Asia and Africa (OED 1991, Murray-Rust and Snellen 1991, Tiffin 1988):

- Many of these systems are characterized by "dual personalities". The first is the nominal system that is assumed to conform to the design and norms of operation, and the second is the reality of its operation. The nominal system is rarely adjusted to the changing economic, social and technical reality.

- Most often the organizational structure is hierarchical, with a top-down mode of operation. Even when the systems are designed to function "managerially", that is with water levels and flows to be adjusted at low levels and some responsiveness to farmers demand, the problems of vertical information transfer and the difficulties of controlling hundreds of adjustable gates are such that these systems are usually only controlled down to a fairly high level. Farmers, either individually or in groups take over control of the outlets and frequently control distribution structures at higher levels than the farm turnouts.

- The dual nature of governmental irrigation systems is fostered by the lack of monitoring and measurement, frequent transfer of engineering staff (with a resulting dependence upon general norms of operation and a lack of familiarity with the local environment), and a pattern of allocation of resources for operation and maintenance that inhibits effective control of the system.

Frequently, other problems with such systems are observed, such as: gaps between actual and expected performance; low water use efficiency; unrealistic designs; designs that are difficult to operate or that cause conflicts among users; inflexible water deliveries; lack of monitoring and field data.

Gap between actual and expected performance

Almost all recent sector studies and evaluation reports note a substantial gap between actual and expected performance. Although part of the problem might be due to overoptimistic and unrealistic expectations when the project was designed or to macroeconomic factors beyond the control of the irrigation agency and the farmers, it is clear that the performance of most irrigation schemes could be much better. A related problem is that performance is defined differently in many reports, making it difficult to compare projects.

Low water-use efficiency

There is increasing awareness that water resources are limited and have to be managed more carefully. Irrigation is by far the largest water user in developing countries, using up to 85 percent of the available water resources. In the future, irrigation will have to share the water resources with industrial and urban water users. The consequences are felt by the farmers who receive only a portion of the flow and the volume of water originally designed. In a recent policy paper, the World Bank stressed the need for comprehensive water management (World Bank, 1993). The immediate conclusion for irrigation is that present water use efficiencies have to increase drastically. Low-water use efficiency leads to unequal and unreliable water distribution and aggravates the tail-end problem. The magnitude of the problem can be observed in Table 1-1.

Table 1-1. Design estimates compared with actual overall irrigation efficiencies

Project	*Design (%)*	*Actual (%)*	*Actual as percent of design*
Sinaloa (Mexico)	52	37	71
Panuco (Mexico)	52	26	50
Doukkala Sprinkler (Morocco)	64	49	77
Doukkala Gravity (Morocco)	50	42	84
Yaqui (Mexico)	43-46	38	85
Coello (Columbia)	-	30	-
Upper Pampanga (Philippines)	58	36	62
Aurora-Penaranda (Philippines)	39	36	92
Lam Pao I (Thailand)	55	28	51
Lam Pao II (Thailand)	58	28	48

Source: Plusquellec et al., 1990.

Unrealistic designs

Often the design and layout of an irrigation scheme fails to consider some basic laws of hydraulics, such as lag time, unsteady nature of water flow, and fluctuations of water level, resulting in poor performance of the scheme. One of the most common design errors is the use of manually operated undershot gates in cross-regulators where almost invariably the function of the structure is to simply maintain a relatively steady upstream water level while passing flow changes on to the next reach. Other design errors traceable to a poor understanding of unsteady flow. All too often the designer assumes the canals will operate well with unsteady flow, but in reality the design prohibits effective operation because it lacks a control strategy, sufficient communications, suitable gate spacings, or other design errors.

Operational problems

Many designs are difficult to manage under real conditions. Operating instructions are often conflicting and sometimes meaningless. Murray-Rust and Snellen (1991), studying the Maneungteung Irrigation Project in Indonesia, observe:

> The system calls for a bi-weekly assessment of demand for every tertiary block, and a readjustment of every gate in the system to meet the changed water distribution plan. This requires a very intensive data collection program and an efficient and effective information management system. Because it is carried out in an environment of unpredictable water availability it becomes almost impossible to achieve even if there were a huge increase in number and skills of field staff.

Another example is the Kirindi Oya irrigation project in Sri Lanka. A simulation model of the present operation of the Right Main Canal showed that it takes up to four days to reach a new steady state after changing the flow at the headworks. In the upper reaches of the canal the steady state is reached soon and the water level fluctuations are low. But in the lower reaches of the canal the water level fluctuations of about one meter occur for up to four days. The results of the simulation model are confirmed by field observations. If the discharge at the headworks is only changed once in a week, steady flow conditions are rarely achieved and frequent fluctuations of water levels and discharge at offtakes are the norm.

There is tremendous room for improvement of operational procedures. An example is the over-emphasis on gate rating tables, which tell operators how much to open a cross-regulator for specific discharges. Some points need to be made in this context:

- Gate rating tables are generally inaccurate. A better design at a bifurcation point is to use a gate to <u>control</u> the flow through an outlet, with a downstream flume or weir to <u>measure</u> that flow or, even better, a flow controller. It is not possible to obtain a constant flow rate into all the branches at a bifurcation point; one of the branches will have to absorb the unavoidable flow fluctuations from upstream.

- For typical cross-regulators in a series in a canal, the use of gate rating tables is meaningless. With upstream control, the flow rate through the structure cannot be controlled by the operator (except for a short duration) because the flow in the canal is determined at the canal inlet. Instructions for cross-regulators with upstream control should deal with maintaining a desired water level, not a desired flow rate.

Water user conflicts

Many authors have observed a strong relationship between design and conflicts over water use. Horst (1987) observes: "It is contended that improvement of management, participation of farmers and action research will remain paper exercises if the design concepts of irrigation canals and structures are not taken into account." Unfortunately, this lesson is seldom heeded. Some irrigation designs guarantee anarchy at the turnouts. When water delivery is erratic, water users lose respect for the rules and regulations governing water usage. These conditions lead to passive water user associations (WUAs) and tremendous damage to distributaries and turnouts. Reports judge the incidence of such damage to be as high as 80 percent, even in some Asian countries with long traditions of irrigation.

Such anarchy is not a necessary part of an irrigation project, nor is it inherent to any culture. Projects with reliable water deliveries in such diverse geographical areas as West Africa, Latin America, and the United States show almost no damage caused by farmers. The authors contend that inadequate design and operation is a much more significant factor in creating conflicts and disorder than the absence of irrigation tradition or social and legal norms.

Lack of flexible water delivery

The cropping pattern in many Asian countries is changing rapidly from a primarily rice culture to more diversified crops. These crops require more frequent irrigations adjusted to different growth stages, soil conditions, rainfall conditions, etc. Water delivery should be responsive to farm needs and preferably on arranged schedule. Increasingly, new water-application methods are used (sprinkler, drip) that require better water control and almost constant supply of water. Traditional on-farm irrigation methods too have a large potential for increased performance and higher water use efficiency. In many cases it is the water supply restraint that forces low performance and inefficient resource utilization on farmers.

Badly conceived and executed pilot projects

Well-conceived pilot projects offer the opportunity to develop new concepts of irrigation, to test equipment and to study the response of users. There are some noteworthy examples of good pilot projects, including the Sidorejo pilot project in Java, Indonesia, the Tepalcatepec modernization project in Mexico, and the Gadigaltar minor irrigation scheme in Madhya Pradesh, India. These projects have benefitted from an integrated, global view of the water delivery and the farming system. Appropriate technology is selected at various levels in these projects to meet well-defined and realistic performance goals.

Unfortunately there are more bad examples than good. Often the pilot project is only half-heartedly supported, or it is placed in the wrong environment. Sometimes the new technology is too sophisticated for the application; in other cases, pilot projects are demonstrations of equipment rather than examples of new water-control concepts, or the new operation mode is completely incompatible with the design of the main system. For example, a pilot project designed to operate with a "demand" delivery schedule may be supplied from the larger project on a rotational basis. Many projects are simply too small or are not given enough time to demonstrate any effect. Often pilot projects are dropped during implementation because host governments and funding agency are not committed to them.

Lack of field data

Systematic monitoring of the process and the result of modernization projects is rare. Seckler describes the problem well.

> "One of the many peculiar facts about irrigation is that the results are rarely monitored. This makes it difficult to learn from experience and leaves the field of irrigation open to the fads and fashions that perpetually

> afflict the development community, causing the donors to chase red herrings that frequently only repeat the mistakes of the past" (Seckler 1993).

If done at all, monitoring is largely confined to economic and social indicators. Remarkably, most of the impact and evaluation studies done in the past ten years have been executed by sociologists and economists, not by engineers. While their participation has contributed to a better understanding of important social and institutional aspects of irrigation performance, design related issues have not received prominence in the evaluation report.

Hypothesis of the Study

Jurriens, Bottrall, et al. (1984) suggest that project evaluators should ask the following questions when faced with any of the above problems: "Would the situation have been better if the design had been different?' If the answer is 'yes,' the second question is: 'What lessons can be drawn for future design?'" They also suggest that "...instead of considering the irrigation system as a given fact and then investigating its operation, one might do better to consider the operational potential as a given fact and then investigate what makes the system so difficult to operate."

The authors fully support this argument. Many of the problems can be solved through a new approach to the design of irrigation systems. The following hypotheses are put forward and discussed in the report:

- A good design increases reliability, equity and flexibility of water delivery to farmers. It reduces conflicts among water users and between water users and the irrigation agency. It leads to lower operational and maintenance costs.

- Extended gravity irrigation schemes with manually operated gates and control structures rarely work, despite all efforts to improve irrigation management and the capacity of staff. The performance is sometimes inferior to systems without adjustable structures. Basically there are two options to improve irrigation performance:(a) simplification through proportional dividers, unadjustable gates and rigid scheduling, or (b) modernization through the application of hydraulic principles, automation, improved communication and decentralization.

- Many failures and problems are caused by a design approach, which pays insufficient attention to operational aspects. A modern design approach is a thought process that starts with the definition of a proper operational plan. The physical configuration and the selection of equipment will be determined by this well-defined operational plan.

- A good design makes maximum use of advanced concepts of hydraulic engineering, agronomic science, irrigation engineering, and social science to produce the simplest and most workable solution. A good design is user friendly and not necessarily synonymous with "high costs," "high maintenance," or "complexity of operation".

- Many designers are not sufficiently aware of new technological options and how different types of equipment interact. Some modernized irrigation projects have failed because of improper choice of control structures, incompatible components, and a design that was not based on realistic operation and maintenance plans. This has created the wrong impression that modern design concepts are not suitable for the environment of developing countries.

- The confusion in the discussion about the performance of irrigation systems is partially caused by the lack of empirical data and the nonrecognition of the different levels of objectives in irrigation systems.

Definition of Modern Design

The terms "modern design" or "modern schemes" are used frequently throughout this report. The authors attach the following definition to these terms:

- Modern irrigation schemes consist of several levels which clearly defined interfaces.

- Each level is technically able to provide reliable, timely and equitable water delivery services to the next lower level. That is, each has the proper types, numbers, and configurations of gates, turnouts, measurement devices, communication systems and other means to control flow rates and water levels as desired.

- An enforceable system is in place that defines the mutual obligations and creates confidence at each level that the next higher level will provide reliable, equitable and timely water delivery service.

- Modern irrigation schemes are responsive to the needs of the end users. Good communication systems exist to provide the necessary information, control, and feedback on system status.

- The hydraulic design of the water-delivery system was created with a well-defined operation plan in mind. The operation plan was established with a clear understanding of the needs of the end users of the water (that is, agronomic and social needs).

- The hydraulic design is robust, in the sense that it will function well in spite of changing channel dimensions, siltation, and communications breakdowns. Automatic devices are used where appropriate to stabilize water levels in unsteady flow conditions.

- Motivated and trained operators are present at all levels of the system. Operating rules for individual operators are well understood and easy to implement.

- A maintenance plan is defined during design, adequately funded through water fees, and strictly implemented.

- There is a recognition of the importance and requirements of agricultural irrigation and the existing social conditions. Engineers do not dictate the terms of water delivery; rather, agricultural and social requirements are understood and satisfied at all levels and at all stages of the design and operation process within overall resource availability.

A modern irrigation design is the result of a thought process that selects the configuration and the physical components in light of a well-defined and realistic operational plan which is based on the service concept. A modern irrigation design is not defined by specific hardware components and control logic, but use of advanced concepts of hydraulic engineering, irrigation engineering, agronomy, and social science should be made to arrive at the most simple and workable solution.

2. DESIGNING FOR OPERATIONAL PERFORMANCE

Visitors to irrigation projects are generally shown the major physical structures, such as the dam, the diversion weir, the main canal, or the control and administrative center. Projects are described in terms of total water supply, size of command area, number of people employed, kilometers of canals, and tonnage of agricultural production. The physical characteristics of the scheme are useful but are only meaningful in the context of the project's objectives. The most important issue, and the one that often receives superficial attention during visits and in project descriptions, is the system's ability to achieve a specific level of operational performance at all levels within the system.

The lack of attention to operational aspects is also true for the design phase. A precondition for high performance, however, is that the design must reflect the objectives and the requirements of future operation. As long as the design of irrigation systems is understood to be a classical engineering task of designing static structures, essential operational questions will not be addressed. The risk is that this approach will result in overall low performance during operation.

Objectives and Performance Criteria of Irrigation Projects

Modern design concepts should be promoted because they improve the performance of irrigation projects. Comparison of the performance of projects of different design is difficult, however, because many different definitions of performance are used. For example, Murray-Rust and Snellen (1991) propose two complementary criteria to measure the overall performance of main system:

- The degree to which the services offered by the main system respond to farmers' needs, within the limitations imposed by national policies and objectives.

- The efficiency with which the irrigation system uses resources in providing these services.

However, Small and Svendsen (1990) suggest three distinct categories of performance measurement:

- Process measures, which relate to a system's internal operations.
- Output measures, which focus on a system's final output.
- Impact measures, which pertain to the effects of a system's outputs on its larger environment.

They have expanded and refined the discussion on the subject in a recent publication (Small and Svendsen, 1992). Abernethy (no date) states that "...the performance of a system is represented by its measured levels of achievement in terms of one, or several, parameters which are chosen as indicators of the system's goals." Suggested indicators of achievement levels include productivity, equity, profitability, quality of life, and sustainability.

Perspectives of Performance

All these definitions contain some important points, and they certainly go well beyond a simplistic definition of performance that relates to pure "irrigation efficiency." However, while these definitions are useful in discussions about irrigation projects at a policy level, it is important to realize that the individuals actually responsible for the final calculations and specifications of the design work may not be familiar with these criteria. Designers of many projects struggle to match a specified water supply with a certain irrigation area and a desired irrigation intensity. As a rule, most designers are immersed in design details. The concepts of "equity, quality of life, and sustainability" never reach their desks. Funding agencies (governments and various donors) are interested in agricultural produce, hectares served, improved income on an aggregate basis, and an anticipated economic rate of return. Evaluators who may visit the projects many years after completion often add extra indicators, such as equity and quality of life. They are not faced with the daily grind of doing the various calculations of design or operations, frequently without the assistance of computers and modern design aids. Evaluators generally deal with senior-level staff in agencies and design organizations[1].

It seems obvious that different objectives are used for irrigation projects as a whole and for each level within a project. These objectives and related performance criteria are often in conflict. In the following the performance criteria will be briefly examined below from the (1) farmer's, (2) operator's, (3) country project manager's and (4) country and evaluator's perspectives. The challenge for the designer is to understand the various perspectives and to minimize conflicts within the system.

The farmer's perspective

The most important performance criteria to the farmer are:

- **Adequacy.** Adequacy is defined as "sufficient volume of water of acceptable quality to irrigate the crops". The adequacy of the water supply has an important bearing on the design of the distribution system and on the need for cooperation among farmers.

- **Reliability.** Abernethy (1991) defines reliability as "the degree to which the irrigation system and its water deliveries conform to the prior expectations of its users." Reliability is absolutely essential to provide proper crop growing conditions and to prevent anarchy among farmers.

- **Timeliness.** Timeliness can be defined (Abernethy 1991) as "the correspondence of water deliveries to crop needs." It also involves aspects of convenience and suitability to the particular on-farm irrigation method used and appropriate frequency and durations for the particular soil-crop-climate combinations involved. Another term frequently used in this context is "flexibility."

---

[1] Only recently have Bos et al. (1993) proposed a comprehensive set of performance indicators which reflect the authors concerns.

- **Flow rate.** Flow rates available to farmers affect the amount of time needed for paddy preparation, the amount of labor needed to irrigate a field, and on-farm irrigation efficiencies.

- **Cost.** Farmers are noted for their reluctance to pay water rates for water from public irrigation schemes. It is interesting to note, however, that the same farmers are often willing to spend considerable amounts (per unit volume of water) to develop groundwater supply. The obvious conclusion is that farmers are willing to pay for water if it is reliable and somewhat flexible.

- **Equity**. If farmers perceive water distribution as inequitable, they will quarrel among themselves or complain to the authorities. If inequities are not corrected, individual farmers may interpret this as a disregard for their rights, and they will take actions to ensure an adequate supply for themselves. Such actions may include damaging gates or turnout structures or even cutting through canal banks.

The field operator's perspective

The operators in the field have different concerns than the farmers. Any attempt to introduce flexible water delivery without physical improvement will likely result in a nightmare for system operators. They typically struggle to manage the irrigation systems as originally designed. Therefore, new schedules and higher performance standards must be accompanied by operational changes and improvements to the infrastructure, such as automation of gates and interim storage. Operator time is consumed by day-to-day problems, such as:

- **Hydraulic difficulties.** Most irrigation systems have not been designed with concern for easy operation. Flow-rate changes made at the head of a canal may take days to reach the end of the canal. Also, stabilization of water levels with manually operated gates may be difficult. Inappropriate designs, poor operating instructions, and lack of communication leave operators in a quandary about what to do.

- **Equipment problems.** Canal linings will fail, gate hoists will break, communications systems fail, and so on.

- **Spills.** Avoidance of spills or canal breaks generally takes precedence over any other matter.

- **Maintenance.** Flow must be maintained at a certain level to minimize deposition of sediments. Silt and weeds must be removed periodically, and all structures need regular maintenance.

In the end, most operators feel fortunate if they can retain the status quo. They need to work hard just to prevent traumatic failures. It is no surprise that they resist new delivery concepts that would increase their workload. From the operators' perspective, the system works best if the farmers or water users groups are prevented from requesting changes in the delivery schedule.

The project manager's perspective

Project managers of public irrigation schemes have their own set of objectives and problems. Politics, finance, and personnel affairs dominate their daily schedules.

- **Personnel issues.** Managers must face all the typical problems associated with the management of personnel: conflict resolution, promotions and transfers, disciplinary actions, payment of salaries and incentives, and so forth. In many cases, managers are forced to hire too many people, as a means of providing employment. The social or legal system may not allow managers to fire incompetent or unnecessary personnel.

- **Politics.** In some projects, managers spend almost all of their time with legislators and regulatory agencies. Typical tasks include trying to retain their "fair share" of water and seeking additional funding for much-needed maintenance and daily operation.

- **Finance.** The funds available to a project manager often originate from the general budget and are not tied to the quality of the irrigation service or the rate of collection of water fees.

- **Environmental concerns.** The first symptoms of trouble in irrigation project design and operation may show up as environmental problems, such as waterlogging and salinity, erosion, or pollution. Managers may spend considerable time providing "Band-aid" solutions to these symptoms.

Project managers of autonomous schemes such as the irrigation districts in the United States, France, or Spain are often able to function similar to managers in the private industry, although most U.S. irrigation projects and districts are local government entities. These managers are generally hired by a board of directors and are directly responsible to them. The financial loop is closed within most irrigation districts, as the revenues generated by the district are spent within the district. This creates a powerful incentive to reduce the annual costs of operation, while maintaining an acceptable level of service.

The country and evaluator's perspectives

At this level, objectives and performance criteria become more comprehensive. Discussions of performance are generally free from such trivial issues as paying bills and preventing water spills. Typical performance criteria may involve:

- **Equity.** Equity of water distribution involves the concepts of fairness, social justice, and redistribution of wealth. The notion of equity invokes many difficult questions. Should there be an equal allocation per area or per household or per person? What about different water rights and different soils? How to account for conveyance losses? It is interesting to note that in traditional, farmer-managed systems, rarely are equal volumetric allocations of water found (Yoder, 1990). This is confirmed by Patil and Datye (1989) who point out that equal treatment of all farms, regardless of soil or crop type, results in poor crop yields.

- **Efficiency.** From the country perspective efficient resource utilization is probably best described by the basin efficiency, which accounts for wasted water that is recovered at the lower end of the system. On-farm, project and basin efficiencies are different. The Nile basin in Egypt, for instance, has an overall basin efficiency of more than 80 percent, although the on-farm application efficiency may only average 30 to 40 percent.

- **Environmental Impact.** Irrigation can have substantial impact on the environment. In addition to concerns about the diminution of river flow in the dry season, apprehension is also caused by the low quality of return flows (both to surface and subsurface water supplies). It is now becoming clear that improved first-time usage of water can mitigate eventual water quality degradation. This makes an argument for improving on-farm irrigation efficiency, rather than for considering the hydrologic basin efficiency, which may be high in some cases as explained above.

- **Employment.** In some areas, it may desirable to maintain high levels of rural employment, though it may be at a low economic level. Project design is different when labor-intensive farming practices are used rather than extensive and mechanized farm operations. Designers often overlook that irrigated agriculture requires substantial additional labor inputs and that this may require farmers, for the first time, to become employers.

- **Profitability.** Financing institutions are primarily interested in "profitability" or "economic rate of return" and financial viability of a project. Planners have assumed that because the farmers receive additional income from a project, they will be willing and able to pay for the operation and maintenance charges. However, it must not be overlooked that the additional labor input requires monetary incentives to be sustainable.

- **Sustainability.** The term has a physical (maintenance), a financial, and an environmental dimension, as discussed above.

- **Quality of life.** A very general objective that is difficult to measure, it involves aspects of food security, regional development, health, employment, and so on.

Where the Planner Stands

Operators and farmers of an irrigation scheme are usually forced to make what exists work. The decisions on the design and layout of the system were made by others, and there is little users and operators can do once a system has been installed.

This situation implies a special responsibility of the designers and planners. Irrigation design is an art as well as a science. There are no guidelines or manuals that are applicable in all cases. New projects or major rehabilitation works are generally financed by the national budget of the country, often with international support. Decision makers are often far removed from the practical considerations of an irrigation project. The design engineer is the key link between the decision makers and the users. A proper operational plan is the instrument that combines the various perspectives and helps to reconcile conflicting expectations.

Elements of an Operation Master Plan

The following are essential components of an operation master plan, which should be defined early in the design phase:

- **Level of service anticipated.** The determining factors for the level of service are discussed in chapters 3 and 4. The plan should address total seasonal allocations, anticipated frequency, rate, and duration of water deliveries to users along with flexibility in changing the schedule or ordering water. In addition, it should define the reliability of service and the consistency of flow rates and pressures during a delivery.

- **Levels of control in major, minor, and watercourse canals and pipes.** The operational objective of each level in a branching distribution system is to provide some degree of service to the next lower level. If a system is designed properly, operators at one level need to concentrate only on proper operation of their level; they do not need to be concerned about details of daily operation for upper and lower levels. IIMI (1989) referred to this concept of "control levels" when stating that "management of the main canal is a hydraulic task that requires no immediate attention to agricultural conditions. It should be well within the capability of professionally trained engineers to operate canal systems to meet the minimum hydraulic conditions required at all structures and gates."

- **Ordering procedure.** The ordering procedure does not just deal with communications for the farmer. An ordering procedure must be defined for all levels of branches within an irrigation system. If one branch needs a change in delivery, who must make the request, in what format, at what time, and to whom? What is then done with that request, and how will the requestor be answered? Is the communication verbal or in writing, and how long will it take to obtain an answer and actually implement the change?

- **Communications.** The details of human, digital, and analog communications must be clearly defined. The specific communications devices and their locations, paths, and reliability will influence the viability of the chosen operation method.

- **Decision support system.** Operation in many projects is supported by computerized decision support systems. Some irrigation agencies are using simulation models to calculate the gate setting from anticipated water deliveries and present system status. The gates will still be operated manually, but the setting will be determined by a mathematical simulation model. A word of caution is appropriate. Simulation models are inherently inaccurate due to calibration and measurement errors. Therefore, the models used must be self-correcting through frequent feed-backs. Experience in the United States (for example, the California Aqueduct) shows that skilled operators refer to the computer-generated gate movements but use their judgment when deciding on actual gate adjustments. Using a predictive model (generally using transfer functions) on a daily basis (Rey

1990) and conveying instructions to individual gate operators may be more complicated than necessary. Gate operators need only very simple instructions (such as to maintain a constant water level upstream of each check structure) to solve the instability problems. Before real-time predictive modeling is implemented, it is important to determine if the problems are simply caused by inadequate or conflicting instructions. For example, if the operators are told to have no spill at the tail end of a canal and to have a constant flow rate out of each offtake, they are being asked to do the impossible on an upstream-controlled canal.

- **Data collection and processing.** The type of decision support system defines the data collection and processing requirements. Levine and Coward (1986) noted that "while procedures for recording and transmitting measurement information frequently exist, these procedures rarely include adequate evaluation of the quality of that information, or the translation of the field data into a form that would permit evaluation of the flow quantities (volume)."

- **Who will do what.** For example, who is responsible for maintenance and operation? The "who" is not only a question of farmers or operators. For various tasks within an irrigation project, it is also a question of operational personnel as opposed to maintenance personnel, and irrigation sector personnel as opposed to agricultural sector staff. Another question is, what are the required qualifications at each level?

- **Operating instructions for specific structures.** Later in the design process specific operating instructions for all structures need to be developed. Wanted and unwanted interactions of the different structures will become apparent only during this process. In irrigation projects throughout the world, operation instructions are often contradictory and sometimes useless. It seems that this arena has tremendous room for improvement.

3. WATER DELIVERY FOR DIFFERENT CLIMATES, CROPS, AND SOILS

Any good operational plan requires a clear understanding and must reflect the climatic conditions, the cropping pattern, and the soil types in the project area. Some aspects that are of importance in the context of irrigation design are briefly reviewed.

Climatic Conditions

Experienced professionals (Berkoff, 1990) have pointed out that the function of irrigation in arid areas differs from that in a humid regions, hence different water-control concepts should be used. The returns from investment in irrigation can be maximized when designing around the local rainfall conditions. It is often overlooked that rainfall in most irrigation areas could be by far the most important source of water.

Arid areas

In well-managed schemes of arid areas, irrigation is more reliable and predictable than rainfall. Therefore, farmers base their decisions to plant on expectations about the availability and reliability of well and surface water. If the water is limited in relation to land, even subsistence farmers will do some "personalized linear programming of their planting" (Berkoff, 1990). They will use the expected water supply for high-value crops on a limited area and either leave the remaining land fallow or plant drought-resistant, low-value crops in the hope of unexpected rainfall. Operation of irrigation schemes in arid regions is usually easier because smaller fluctuations in water demand require fewer provisions to regulate unsteady flow. In addition, farmer organizations are generally stronger in an arid environment, and severe social sanctions are imposed on free riders. The implication for the designer is that the reliability of water supply should be the overriding consideration, followed by equity concerns. As a first step, sophisticated control structures are of lesser importance as long as the water distribution is fairly effective.

Humid areas

In humid areas, planting decisions (date and area) are dominated by expected rainfall rather than based on expected irrigation supply. Usually the total agricultural area is planted with high-value crops since farms are usually small and land is scarce. This situation poses several problems for irrigation designers and operators. Although farmers rely basically on rainfall and the irrigation system provides only some supplementary water under normal conditions, the system must still be able to satisfy the total evapotranspiration requirement (plus inefficiencies) in periods of drought. When the rain does not arrive in time, everybody tries desperately to save his crop, and the management task of the system operator becomes formidable. The problem is exacerbated by weak organization at the water-user level because the irrigation system is often not considered an essential part of the traditional social structure.

Seasonal variation

In many monsoon regions farmers are growing increasingly diversified crops (for example, vegetables and root crops) rather than rice in the dry season. These crops require less water but call for a different irrigation technique. The problem is that continuous flow to outlets and field-to-field water distribution, common for paddy cultivation, doesn't work for diversified cropping. Better water control and improved distribution systems are, therefore, required.

In the wet season the rainfall pattern directly influences water demand and supply. Heavy rainfall in one area might reduce the demand and close parts of the system temporarily. Rivers might rise unexpectedly, causing the system operator to close the intake. The storage capacity of dams can be better utilized if deliveries can be shut off when it rains. Valera and Desa (1991) state that in a project in Malaysia, the rainfall effectiveness is only about 30 percent, since farmers cannot shut the water off during rainfall. Irrigation schemes in humid areas need to be responsive to conditions, not preprogrammed, if water use is to be maximized.

Rice cultivation presents special problems for irrigation designers. The monsoon season may last only a few months, and farmers need water for nurseries, soaking of land, and puddling. Canals need large capacities to make sure that all the land preparation can be completed before the onset of the monsoon season. Several weeks of delay in planting means several weeks of extra irrigation needed from the project after the monsoon has stopped, resulting in an ineffective utilization of rainfall. Also, a shift from rice transplanting to direct seeding (broadcasting) requires better water control within the distribution system. Direct-seeded rice is particularly sensitive to water depth, especially if pregerminated seeds are used in manual wet seeding. In addition, different farmers using different planting techniques require different amounts of time and water at different times (OED 1991).

Conclusions

The following points summarize design considerations related to climatic conditions:

- In humid areas water-delivery systems have to be designed for frequent flow changes and rapid closure.

- Projects in humid areas need to be sized for peak evapotranspiration requirements during drought periods.

- Rice projects need large turnout and canal capacities to accelerate land preparation before monsoon season.

- Daily or weekly predictions of water demand from climatic data are fairly useless for operation in humid areas since local rainfall can change water demand rapidly. These predictions may help to estimate gross demand of the main system in arid areas with low rainfall probability.

There has been some debate on whether modern water control concepts are equally applicable in arid and humid regions. The authors argue that there is sufficient evidence that the application of modern concepts is not limited to arid regions. The experiences with irrigation schemes in different climatic zones, as presented in Part II, demonstrate that substantial improvements of irrigation performance can be achieved through modern water control methods.

Crops

There are inherent differences among crops in relation to drought resistance, needs for water at particular growth stages, root depths, required frequency of irrigation, and relative costs of water as compared with the costs of other input. This fact has profound implications on the design and operation of the main system. Since the cropping pattern will change frequently in response to market forces through the life of the project, sufficient flexibility is required to meet changing crop water demands.

Rice

The previous sections covered some special features of rice paddy cultivation. In addition, the delivery schedule during the growth period (not land preparation) need to be considered. During the growth period, farmers traditionally prefer a continuous supply of water, which enables them to maintain the desired water levels in the paddy fields to suppress weed growth. Attempts to introduce a rotational schedule in rice schemes, such as in Madagascar, have failed. Farmers ignored the new schedule because of the convenience of continuous flow.

Other grain crops

Most grain crops are relatively insensitive to occasional water stress. Some yield is usually obtained, in particular if the crops are planted at the end of the wet season. During early stages of growth grain crops are very sensitive to water stress, but proper timing and use of late rains and residual soil moisture helps these crops to survive the critical period. In later growth stages there is only one other critical period -- pollination. If corn wilts for only two days during the pollination stage, grain yield can be reduced by as much as 20 percent. Grain crops can be grown very successfully on an irrigation schedule of two to three weeks' interval. The intervals can be adjusted to varying evapotranspiration and rooting depths.

Trees and grapes

Mature trees have a deep and nonexpanding root zone. They do well if irrigated by traditional surface irrigation at fairly long intervals. For maximum production and quality of fruit, however, modern farmers are using more and more high-frequency irrigation such as drip. Drip systems however, use only a portion of the total soil field capacity. Unintentional interruption of water supply is, therefore, more serious than with surface irrigation. The drip systems require a continuous flow, implying very reliable water delivery. An additional consideration with pressurized irrigation systems is that pumps can break down or shut off unexpectedly because of mechanical problems or power failures. The main system must be sufficiently flexible to accommodate these fluctuations. Reliability of water delivery is critical if modern, pressurized on-farm irrigation systems are used.

Vegetables

Vegetables are characterized by their high commercial value but sensitivity to water stress and shallow root zones. Vegetables are often grown on small fields having wide variations of planting and harvest dates. Water stress at almost any stage of growth can impair the desired quality (appearance, fiber content, firmness, and so on) and the value of the whole crop, therefore water supply must be extremely reliable. Rotation schedules are not practical for vegetable production.

Sugar cane, alfalfa, and pasture

The vegetative growth of sugarcane, alfalfa, and pasture is completely harvested. There is no critical growth stage after the plant is established. These crops have deep root zones and yield does not drop dramatically as a result of occasional water stress. A ten percent reduction in water use will result only in a small decrease in the production of green matter. The crops are the least sensitive to irregular water deliveries and long irrigation intervals.

Other considerations

The discussion of the "dry-footed" crops above concentrated on the effects of scarcity of water. Equally important is the effect of too much water. All crops suffer yield decline or fail completely if exposed for prolonged periods to excess water. Therefore, farmers should not be obliged through faulty system design to accept more water than required. Poor water control in one part of the scheme may also result in high water table and salinity problems in other areas. In the San Joaquin Valley of California this has been a major problem. In some places farmers are no longer able to grow salt sensitive crops such as tomatoes and lettuce. These problems are often not the result of incorrect irrigation by the affected grower but the result of bad water management and lack of drainage occurring elsewhere. In the end, good on-farm irrigation practices for sustainable irrigated agriculture depends upon controllable and reliable water deliveries.

Soils

More than anything else, differences in soil conditions influence on-farm irrigation requirements. Crop evapotranspiration is equal in well-managed fields on sandy and clay soils. There are major differences, however, between these soils with regard to optimum irrigation frequency, flow rate, and duration.

For upland (dry foot) crops, both the water holding capacity and the intake rate of a soil are important. Sandy soils must be irrigated frequently (due to low available-water-holding capacities) and with high flow rates and short duration (because of high intake rates). Loam and clay soils have high field capacity and low infiltration rates, thus requiring small flow but longer duration and longer intervals. Cracking clay soils present a special problem: if the irrigations are not frequent enough, large cracks will develop and infiltrate large quantities of water. Sometimes the large cracks act as channels to sandy subsurface soil layers, and large quantities of water are lost for consumptive use.

No one irrigation schedule (rate, duration, interval) is optimal for all soil types. Recognition of the importance of customizing water deliveries to different soil types is a major reason that modern irrigation schemes provide as much flexibility in water delivery as possible, rather than forcing the users to adapt to a rotation of a specific flow rate, duration, and frequency. For optimum water-use efficiency, irrigation must be customized in timing and application rate to match the specific soil and crop conditions.

4. SCHEDULING OF WATER DELIVERIES

Irrigation systems should be designed with a certain mode of operation in mind. The operation of an irrigation system, as described in this report, consists of acting upon two primary decisions. First, the scheduling of water deliveries. This can described as frequency, rate, and duration of water deliveries at all levels within an irrigation conveyance system. Second, the determination of the interactive movement of various control structures to accomplish the desired schedule.

This chapter deals with the scheduling of water deliveries. Primary topics include:

- Brief descriptions of the common types of water delivery schedules used both at the farmer turnout and at lateral turnouts.
- Basic design requirements for various types of schedules.
- The difference of deliveries in terms of "flow rate" or "volume."
- A case for abandoning the hope that centralized scheduling of on-farm water deliveries will optimize performance.

Definitions and concepts

Terms such as "rotation," "arranged schedule," "limited rate demand," and "centralized scheduling", as found in the literature (Clemmens, 1987) and described later in this chapter, almost always refer to the scheduling of water deliveries to the farmer or to the final outlet. Those terms are equally applicable, however, to the scheduling of deliveries to all levels within a distribution system.

Water deliveries are made to each branch of a distribution system from the upstream branch. Main canals make deliveries to secondary canals, secondaries to tertiaries, and so on. Improved water control must begin with the main canals. Unless the unsteady flow in the main canals can be easily controlled, the flows in secondary canals will be erratic. Some exceptions exist, such as when there is significant buffer reservoir capacity at lower levels in the system or when the main system is deliberately managed with a variable spill at the tail end.

The majority of irrigation projects in the world have some type of rotation delivery system. In areas with advanced technologies, rotation schedules have largely been replaced by more flexible deliveries. In California, for example, less than 15 percent of the surface-supplied irrigation acreage still has some form of rotation schedule, and that rotation schedule is generally flexible for flow rate.

The water-delivery schedules are defined by the following characteristics:

- Frequency (How often the water arrives?);
- Rate (How much water flows per unit of time?);
- Duration (How long is the flow rate delivered?);
- Timeliness (Does the water arrive when the crop needs it or the farmer can use it?).

There are three important points to consider in scheduling water-delivery. First, the fact that an irrigation project has adopted a certain schedule of water deliveries does not imply that the frequency, rate, duration, and timeliness are equitable or reliable. Those variables may have uncontrolled space and time differences between outlets. It is a common misunderstanding that a theoretically rigid and simple water delivery schedule will necessarily be reliable and equitable. A recent survey of 43 canals in Punjab, Pakistan, indicated that 50 percent of the users draw 53 percent more water than assumed in the design (Bhutta, 1989).

Second, a water delivery schedule does not necessarily imply a specific design. A fairly rigid schedule of water deliveries to the ultimate turnouts may use modern irrigation hardware and computerized decision support systems to make the deliveries equitable and reliable. A primary advantage of using modern design concepts and equipment is that the operators can choose the degree of flexibility offered at various levels within a system. In Morocco, for example, many projects are based upon the same design principles, but there are variations in the delivery schedules.

A third and major point is that a schedule defined at the central office may never materialize in the field as intended. For instance, fixed schedules are published and prominently displayed on boards in many schemes, but farmers ignore those schedules. It takes more than wishful thinking and a paper schedule to produce results. Appropriate design, communication facilities and responsive management are required. Murray-Rust (1990) states that reality is too often ignored, and "management continues as if the tasks can be (and are) undertaken."

Location of Decision Making

Clemmens (1987) describes various water delivery schedules, with their intended frequency, rate, and duration. Traditional delivery schedules (Warabandi and most rotational schedules) have very little or no flexibility built into them. They do not attempt to match water deliveries to crop needs; rather, the farmer must match his planting and management to the water deliveries. The stated objective of these rigid schedules is to obtain equity through simplicity, although poor design, maintenance and operational problems may not achieve the objective. The decisions are made at the central office; even if there is feedback from the field, the delivery system is rarely capable of responding or changing rapidly.

New irrigation projects are generally built with the stated objective of delivering water according to crop water requirements. This objective implies a delivery schedule with more flexibility than a simple rotation. As a minimum, the frequency (and sometimes the duration) of irrigations must be adjusted throughout the growing season in order to properly refill the soil moisture reservoir as the plant matures and the weather changes.

There are two fundamentally divergent opinions among irrigation experts designing an effective delivery system with flexibility needed to match crop water requirements:

- **A top-down approach**, in which crop water requirements are computed by the project center and the allocations to users are made via centralized decisions.

These decisions may be made with the help of computer models (Merkley and Walker, 1991).

- **A bottom-up approach**, in which the delivery system is somewhat responsive to the requirements of the users themselves (or groups of users), and the project center does not attempt to predict individual field water requirements. The center, however, may make use of predictive techniques to anticipate gross flow-rate changes in the main and secondary canals.

The reason to stress the existence of and differences between the two approaches is that projects with stated objectives of "providing water based upon crop requirements" may have widely varying design and operation assumptions. Therefore, the existence of such an objective is not by itself evidence of the use of modern and appropriate design.

Review and Discussion of Water-Delivery Schedules

There are different types of water-delivery schedules: rotation schedule, centralized scheduling, arranged schedule and limited rate demand.

Rotation schedules

Rotation schedules provide water with no flexibility in frequency, rate, or duration. In the simplest form of these schedules, a fixed flow rate arrives at a certain point on a calendar basis (for example, once every one or two weeks) for a fixed duration. In practice, there are many variations of rotation schedules. Some schemes vary the frequency or duration of delivery a few times during the season in an attempt to compensate for different evapotranspiration rates. Most rotation schedules always deliver the same flow rate. This flow rate is rotated among various users (one at a time) on a lateral, distributary, or tertiary canal until every user has received water for his turn.

The warabandi rotation schedule is common in northwest India and Pakistan. It is well described by Berkoff (1990). The warabandi is essentially a rigid water spreading schedule intended to be fairly equitable. Although the warabandi is a schedule, certain types of structures such as flow dividers and nonadjustable outlets, have been associated with it. There is very little or no pretext of flexibility. The water management concept of the La Joya irrigation district in Peru (Part 2, Peru) presents some similarities with the warabandi system of India and Pakistan. It also features on equal distribution between sectors and among farmers through rotational schedules and non-adjustable structures. However, accurate flow dividers, made possible by the large head in the system, are an important feature that secure the equality of water distribution.

Three important questions dominate the design of warabandi systems of India and Pakistan. These are:

- **What is the water duty?** Water duty is defined as the flow rate per gross area served. The concept of volumetric water delivery, as is found with modern irrigation projects, is absent.

- **How to minimize human intervention?.** The standard solution is to design a system that is virtually nonresponsive and nonadjustable (even if adjustments are needed). This design characteristic has caused problems when the warabandi is instituted in high-rainfall areas; even when it rains it is difficult to close the water because this will disturb the rotation.

- **How to minimize the sedimentation problem?** The solution is to operate the canals in an on/off mode at near full capacity and designing the canals for non-scouring, non-depositing velocity. It is recognized that the siltation problems are a strong argument in favor of the warabandi system and against the introduction of unsteady flow regimes. Further discussion on the subject is presented in Section 7.1 of the report.

In a rotational system every irrigator knows exactly when he should receive water, and all irrigators understand and share the risks of underirrigation. The warabandi system is used successfully in areas where the crops are grains or other deep-rooted crops, the climate is arid or semiarid, the irrigation is extensive; that is, there is not enough water to provide full irrigation for all the acreage, and in many areas fresh groundwater is available to provide a flexible water supply.

IIMI (1989) describes several types of on-off strategies used in rotation schedules during periods of low flow or low water demand. No special logic could be identified that governed the application of one concept or another. The four main types observed were:

- Rotation between sub-groups within a tertiary block to accommodate reduced flows in a tertiary canal that does not allow all subgroups to irrigate simultaneously. This procedure requires farmer cooperation, but minimal input from the gatekeeper.

- Rotation between tertiary blocks along a secondary canal that permits a tertiary block to receive its full discharge for a shorter time or at longer intervals. This requires no farmer participation, but does require active involvement of gatekeepers.

- Rotation between secondary canals, meaning all tertiary blocks along a reach of canal receive water simultaneously for a limited period.

- Rotation between successive seasons, where certain areas receive water every other dry season but receive enough water to irrigate the entire area during that season. This approach was tried successfully in Way Jepara, Indonesia (Murray-Rust and Snellen 1991) for several years. It requires mutual trust and strong social coherence. The method is problematic when the water supply conditions are changing, for example because of interventions in the catchment.

Another method, found in tank irrigation schemes in Sri Lanka, is the sharing of irrigated land, the size of which is determined by water availability. This method requires strong water user associations.

The advocates of rotational schedules emphasize their simplicity and implied equity of water deliveries. These advantages are supposed to compensate for the lack of flexibility. However, many irrigation projects in Asia with rotation schedules are designed with proportional-type on-off structures. Such a design may be equitable at full flow and with regular maintenance. In practice the deliveries are not equitable because there is always siltation or erosion in the canals. A canal designed for full-flow conditions gives very unequal water distribution at lower flow rates, even if these are only 10 to 15 percent lower than anticipated. A good understanding of unsteady flow supports this argument even without verification in the field.

Central scheduling

Central scheduling is a "top-down" scheduling procedure. It is used in most Asian countries. The central command uses wisdom, assumptions, and possibly unsteady flow computer models to determine the schedules for water delivery at all points within the system. The schedule is not responsive; farmers do not order water. The central management determines when and how much water is to be delivered. In one sense, it is extremely simple because there is no need for farmer input. Justifications of this control logic often include the following:

- Water is a scarce resource that has to be distributed equally over as wide an area as possible. In the absence of market mechanisms, such as price, water has to be allocated through central decision making.

- It is difficult, if not impossible, to interact with a large number of farmers or farmer organizations.

- The central command has more insight than the farmers in matters such as irrigation scheduling, cropping patterns, and plant water requirements.

Central scheduling procedure is implemented in the form of variable releases at the dam or diversion site on instruction of the central command, accompanied by instructions to operators of cross-regulators within the system. There is, however, an important difference with respect to flexibility: Many recently built "flexible" delivery system designs are flexible only in the sense that the central command may change its mind about water requirements and releases. The truly flexible systems in North Africa and the Americas, however, are flexible in the sense that the central command can respond (within limits) to requirements that originate at lower levels.

This difference in the definition of "flexibility" may become clearer when analyzing the following quotation:

> "In practice most irrigation engineering designs are based on a large number of physical and economical parameters: crops, topography, hydrology, climate, hydraulics, cost/benefit analysis, etc. The aim is to meet the varying irrigation requirements for a predetermined assumed cropping calendar with the highest possible efficiency and operational flexibility....Systems with high flexibility and (assumed) high efficiencies require sophisticated (movable-adjustable) structures and measuring

devices together with considerable numbers of highly qualified operating staff" (Horst 1987).

Reading the quotation does not immediately inform the reader whether or not the "flexibility" is determined at the center or by feedback from the end users. Both methods, the responsive and dictated, could be described as having "the aim to meet the varying irrigation water requirements." The basic difference between a responsive and a dictated flexibility lies in the few words "predetermined assumed cropping calendar". This implies that the quotation referred to a schedule with a "dictated" flexibility.

The authors agree that systems with a centrally planned schemes with large number of manually operated control structures (gates) are indeed difficult to operate. Therefore, they recommend a different, "responsive," approach to design and operation. An examination of many modern irrigation systems (for example in Europe, the United States, and Latin America) shows that it is entirely possible to have a very effective flexible system without knowing or imposing the cropping pattern.

A flexible operation must respond to local conditions, and therefore operational decisions should be decentralized as much as possible. For example, as long as a main canal is operated to provide water to the secondary canals (expressed in terms of flow rates needed for various durations, as requested by the operators of the secondary canals), the operators of the main canal don't need to understand the details of how the secondary operators arrived at their particular request.

In some cases central scheduling may be justified, especially when water is scarce. The use of computer models, remote monitoring and good communication can indeed provide substantial improvement. Unfortunately, more often the top-down decisions are based upon incorrect information (both hydraulic and agronomic) and are not passed down to the system operators in a timely manner. Also, the control structures in the system do not provide the flexibility and accuracy to implement the required changes.

Arranged schedule

The most common water-delivery schedule to farms using modern on-farm irrigation technology is the arranged schedule. Water requests are made for a specific date, a specific flow and a specific duration. Sometimes restrictions on the maximum flow apply. The schedule requires good communication between the water user and the irrigation agency, and a flexible delivery system. Water requests are usually required 24 to 48 hours in advance to arrange delivery. In most systems, the farmer is not allowed to change the prearranged flow rate or duration for that event, although the farmer may order another combination of flow rate and duration for the next irrigation. In more flexible systems, the farmer can personally open and close the turnout (at the prearranged time) and may be allowed to shut off early if some advance notice is provided to the irrigation project. Arranged schedules in a water-short situation require the allocation of volumetric water quotas to the individual farmers or farmer groups.

Arranged schedules are often linked to volumetric water charges. In that case individual field turnouts must be metered, because each field may receive a different volume of water. Arranged schedules are the predominate delivery schedule in California and in most projects in Mexico, Colombia, and North Africa. Arranged schedules are not limited to large fields. The schemes in the above countries have a wide range of field sizes. In California, for example, some districts have an average field size of close to 5 hectares; in other districts the average field size is 50 hectares.

Limited rate demand

Limited rate demand refers to the schedule provided to urban homeowners for domestic water supply. There is no need to make any request for water; it is available at all times. The maximum flow rate is limited, however, by the size of the line servicing individual homes.

In agriculture, there are very few such systems, except private wells. This schedule needs very flexible water-delivery systems that can respond automatically to starts and stops of turnout flow rates. In some cases, the project management may require an advance notice before the water is turned on (but not before shutting it off), so that the project authorities can make sure there is sufficient capacity in the distribution system. Two examples of the limited rate demand schedule are the Canal de Provence in southern France and the Tranquillity Irrigation District in the San Joaquin Valley of California. In Provence, the water is distributed through low pressure pipes to hydrants at the farm boundaries. In Tranquillity, the system is very old, with level-top canals, and almost all of the on-farm irrigation is done with furrows.

It is unrealistic to expect that most existing irrigation systems can be modified in the near future to provide a limited rate demand schedule; in most cases the costs are too high. Furthermore, most farmers and small groups of farmers can work well with a flexible arranged schedule, especially if they are given some latitude regarding early shutoffs.

Intensive Compared with Extensive Irrigation

A key issue, the planner must address, is the intensity of irrigation. The question is whether the project should be designed to meet the peak water requirement of only a part or of the entire project area. The problem transcends the scope of engineering and is strongly influenced by economic and social considerations. Extensive systems are more costly to build and to maintain. However, the irrigation service provided to end users is often limited to only two to three irrigations per season. Therefore, the increase of crop production may not sufficient to pay for the cost of investment and operation of extensive systems. In reality overextended systems often shrink as government withdraws from operation and maintenance. On the other hand, extensive systems often have less problems with waterlogging and salinity, and provide incentives for conjunctive use of ground and surface water. If extensive systems are built for social reasons then the question of cost recovery and maintenance costs need to be addressed at the beginning and should be considered, at least partially, as a permanent social subsidy. OED (1993) proposes specific studies to decide on the optimal command area to be irrigated from a given source.

Smallholders agriculture and conjunctive use

Traditionally, the irrigation systems in northern India and Pakistan have been built for drought protection. Water was spread over a very wide area. The volume delivered was only about one-third of the total requirements. Proportional distribution systems and rotational delivery (warabandi) were widely used. Is there any need to modernize these schemes?

Berkoff (1990) believes that in those areas of northwest India that are underlain by fresh, shallow groundwater, there is no need to make the water available on an arranged basis. His assumption is that making water delivery responsive to users requests would make the system unmanageable. He states that:

> The first step must be to gain control since otherwise no management is possible. To achieve control, trade-offs may be necessary between the ability to respond and manageability. Since water is scarce relative to land, the objective must be the conjunctive use of all water resources, defined here as the optimum use of surface water, groundwater and rainfall under conditions where the only significant practical public intervention is the way that water is distributed in time and space in the surface system. Management is thus the key, in large surface systems, to the use not only of surface water but also of rainfall (the largest resource in quantitative terms) and groundwater (the most valuable resource, given the control it allows the farmer). It may also be an important contributing factor to resolving drainage problems which in the east (of India) can be the dominant constraint and which in the west must be resolved if irrigated agriculture is to be sustained over large areas underlain by saline groundwater."

In other words, Berkoff argues that flexibility will be provided through groundwater use, whereas surface water will be strictly controlled and managed by a central authority on a rotational basis. This idea has merits for the specific combination of arid environment, fresh shallow groundwater, and smallholder farming. However, Berkhoff might have overlooked that with better technology and organization the management dilemma can be solved.

Smallholders agriculture and storage schemes

In areas without access to shallow groundwater it may be possible to provide arranged schedules to individual fields, even with limited water supply and smallholder agriculture. The requirements are:

- Upstream storage capacity of sufficient size.

- Reliable deliveries on the basis of fixed annual or seasonal water allocations agreed upon by the irrigation agency and the water user association.

- An appropriate water-control system at the main and secondary levels that allows flexibility in frequency, rate, and duration at these upper levels. The final watercourses may be operated with less flexibility, but still on an arranged schedule.

- An effective water user association, with well-defined rules and a means of enforcing those rules. Employees of the farmers must actually make water deliveries.

- Deliveries systems that do not require much cooperation between farmers and prevent interventions. This presents a strong argument in favor for delivery through low-pressure pipes or lined watercourses.

A good example of such an arrangement is the Datta Water Distribution Cooperative Society of the Mula Irrigation Project, Ahmednagar in Maharashtra (Lele and Patil, 1993).

Commercial agriculture

Conditions are different in commercial agriculture, where fields are large enough to justify accurate water measuring devices. In the western United States there are many irrigation schemes that provide water on an arranged schedule even though the water supply is limited. Operation is based on the following principles:

- The total volume of water, likely to be available over the season to the district, is known by all and distributed to individual farmers in accordance with established rules and water rights. The risks associated with inaccurate estimates are shared equally by all.

- Water deliveries are volumetrically metered.

- Farmers decide on all aspects of irrigation within the irrigation districts, in particular for not exceeding their quota.

- High flexibility of water delivery allows farmers to apply water in accordance with the needs and constraints of farm operations.

- Groundwater is used extensively to supplement surface water supplies in drought years. The aquifers are artificially recharged during wet years or whenever surplus water is available. Almost all farmers operate individual wells.

Design Concepts for Final Distribution

Two different concepts influence the design of the final distribution system: (1) flow rate and (2) volume. Extremely important, yet sometimes subtle, differences exist between these two concepts. The implications for design and management are substantial.

Flow rate

Designing for a flow rate is the predominant design concept in Asia, the flow rate forms the basis for most systems using a rotation water delivery schedule. The concept assumes essentially steady flow conditions and addresses only one aspect of irrigation design, that of canal capacity under maximum flow conditions. However, unsteady flow conditions always occur, regardless of what the initial design assumptions are. The design does not adequately address the following important issues:

- Lag time, that is the delay between making a flow-rate change at one part in a canal and the arrival of that change at a point downstream.
- Fluctuations of flow at turnouts.
- Operational storage.
- A control strategy, appropriate check structures and communication systems to allow for control of unsteady flow.

The concept of designing and managing an irrigation project for a certain flow rate (rather than volumetric deliveries) is one of water spreading. The farmer has no choice but to adapt to the water delivery system, rather than vice versa. This may be a justified decision under certain conditions, especially with run-of-the-river schemes (Berkoff,1990), but more often it reflects a design concept with little concern for agronomic and on-farm irrigation requirements.

The term "water duty" is frequently used in connection with the flow rate. The water duty is a design criteria which relates an area to a flow rate (for example 400 acres/cusec in India, or perhaps 0.4 liters/sec/ha in other countries). Of course, all irrigation conveyance systems must be designed for some maximum capacity. But for operation the water duty is only important a few times of the year, such as during land preparation for rice cultivation and during periods of peak evapotranspiration. Some designers, however, seem to assume that the scheme is operated at the water duty throughout the year. Therefore, their design tends to ignore structures and capacities for proper control operation at less than full capacity.[2]

Volumetric supply

A water-distribution system intended to provide water volumes to individuals or groups of farmers at appropriate times, durations, and flow rates is approached from an entirely different perspective. For maximum on-farm effectiveness, farmers must apply appropriate volumes at frequencies and durations to match their specific labor-soil-crop-irrigation system. As mentioned earlier, this does not mean that farmers need to have water available on demand. An arranged schedule or a modified rotation schedule can often be quite sufficient. The major difference is that by designing for volumetric deliveries a designer must allow for a relatively larger capacity of the canals in the lower parts of the system, unsteady flow conditions in the main canals and two-way communication between farmers and operators. Volumetric delivery generally implies some type of water billing based upon volume delivered. In most cases, volume is determined by recording

[2] In the warabandi system tertiary canals are indeed only operated at full capacity, even in water short periods or during times of low water demand. They are therefore designed without regulation structures. The drawback of this practice are unacceptable long intervals between irrigations and unavoidable changes of the design water level, as discussed in Chapter 4, Central scheduling.

a certain discharge for a period of time. In more advanced systems, totalizing flow meters are used.

The different approaches of determining the canal capacity lead to vastly different results as can be seen in figure 1 below. Relative canal capacity (flow rate per area served) in typical North African, French, and U.S. schemes increases as one moves downstream. These designs are based on the volume concept and recognize the need for flexibility. They are recommended for responsive systems. In most Asian projects, which are designed on the water duty, the relative canal capacity decreases as one moves downstream, since only seepage losses were considered.

Figure 1 Design capacity of canals and pipelines vs. area served in different design concepts

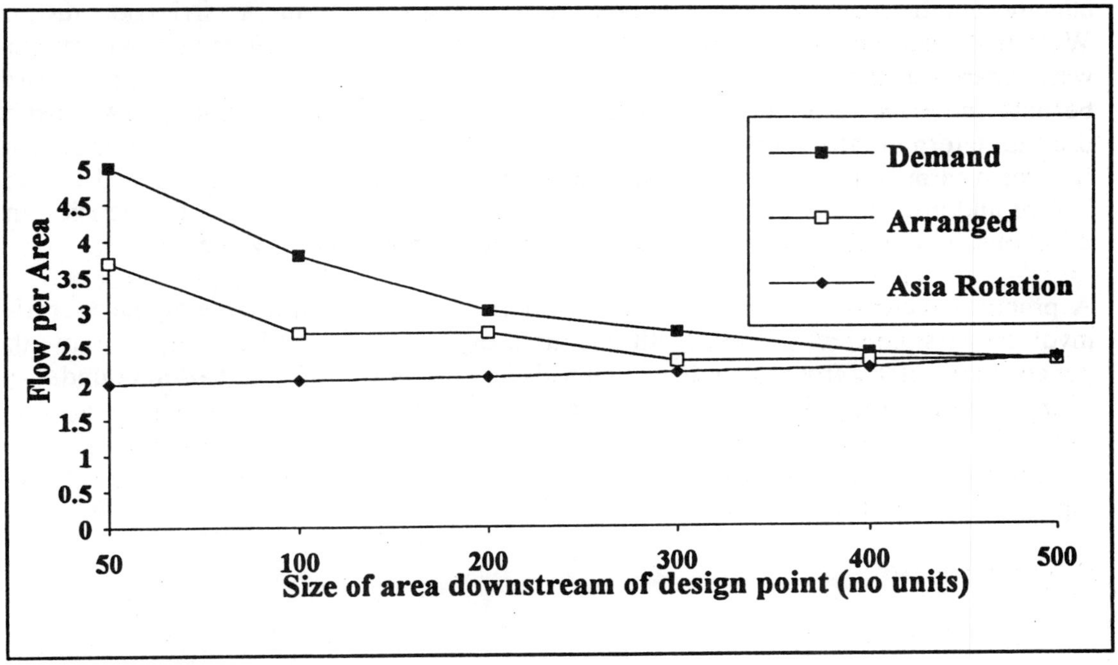

Source: Charles Burt

Note that the values in Figure 1 are relative, but a flow rate/area of 2.0 represents the approximate continuous, unstressed water requirement, including losses, for a relative area of 50 (no units).

Determination of on-farm requirements

Two different ideas need to be combined in modern on-farm irrigation scheduling:

- Determination of the rate at which the soil moisture is depleted by the crop.
- Determination of the proper frequency, rate, and duration of the actual water delivery to the field.

Most of the literature on irrigation scheduling considers only the first item. Numerous books, papers, and research reports discuss methods to determine the evapotranspiration (ET) and the soil moisture balance. In practice, operational plans which depend upon these factors as the only basis for daily water deliveries perform poorly. Why is this so?

The answer lies in the neglect of the second item. Skilled irrigators know that in reality the irrigation method itself may be more important in deciding the proper irrigation frequency and duration. Knowledge of factors such as water application rate, soil water intake rates, and labor constraints, in conjunction with knowledge of soil moisture depletion, is needed to determine the proper irrigation schedule. These conditions change frequently, they also vary from farm to farm and field to field.

In theory, an operation plan can be developed in which centralized computers monitor climatic conditions and estimate evapotranspiration throughout an irrigation district. Water can then be released into the project based upon that estimate. However, this water management by the central authority rarely if ever matches the on-farm needs because the estimate of evapotranspiration is only one piece of information that must be used to determine the proper frequency, rate, and duration of irrigation of an individual field or a farm. In practice, even simple soil moisture observations by trained farmers before and after irrigations can be used to determine proper irrigation frequencies and durations, without ever having to calculate an evapotranspiration value.

A practical technique for controlling irrigation in advanced and developing countries involves a simple steel probe. With adequate experience it can be used to follow the penetrating wetting front when irrigating with furrows or sprinkler or to detect under- or overirrigation when using border strips or basin irrigation.

5. WATER-CONTROL METHODS

This chapter introduces some basic concepts needed to select the water-control method and hydraulic equipment during the design process. It is not intended to present a comprehensive description of the different control methods and hydraulic equipment used in canal regulation and piped distribution. The reader is referred to a recent International Commission on Irrigation and Drainage publication (Goussard, 1993) and for a general introduction to an audiovisual program produced by the World Bank (Plusquellec, 1988). Part II of this paper includes a brief description of the application of different water control methods in a variety of projects throughout the world. Annex 1 summarizes characteristics of the major control methods and their accompanying structures. Advantages and disadvantages are listed, as well as application and design notes.

Three elements together define the water-control method: the control strategy, the hydraulic equipment, and the layout of the distribution system. Three facts are particular important:

- There is no such thing as "best" water-control strategy. The selection of the strategy is the result of a design process as further explained in the next chapter.
- The same hydraulic equipment can be used for different control methods. It is important to fully understand the interaction of different structures under unsteady flow conditions.
- The layout and sizing of the distribution system have profound influence on the water control method. Appropriate design greatly increases an irrigation system's capacity to provide flexible service.

Water Control Strategies

Several control strategies are being used in irrigation schemes throughout the world:

- Proportional control
- Adjustable flow-rate control
- Upstream control
- Downstream control
- Remote monitoring
- Remote control

The list is not complete because each method includes several subclassifications. For example, proportional control may be nonadjustable or adjustable. Downstream control may be centralized or local, and local downstream control may be on sloping or level-top canals or pipelines. The term "upstream control" describes a control method that maintains a constant water level upstream of a check structure or, less frequently, a method that maintains a constant flow through the check structure.

No control concept is optimal for all conditions. For example, under certain conditions it is possible to provide flexible water deliveries with upstream control (assuming that some tail-end spill is allowed), although upstream control is usually associated with rigid, top-down water delivery. Examples include the Imperial Irrigation District in California and the Coello scheme in Colombia. A detailed discussion of the different strategies would exceed the scope of this paper.

Several points are worth mentioning, however, because they are frequently overlooked. First, selection of appropriate water-control methods can have two very significant effects on day-to-day operation. Certain control methods eliminate the need for advance scheduling of water deliveries, the need to know exactly the flows at various locations in canals, the determination of lag time, and the estimations of seepage and canal conveyance losses. Downstream control methods such as constant volume control on level-top canals (BIVAL), downstream control on sloping canals (EL-FLO, CARDD) or dynamic regulation (Canal de Provence) greatly simplify canal operations. Moreover, appropriate equipment selection will automatically stabilize flows and water levels, thus eliminating the burden on the operators of having to adjust constantly the gate positions.

Second, canal regulation can be considerably improved even without investing in new equipment and communication systems by introducing suitable control strategies. Upstream control often requires the operator at each control structure to maintain a certain flow through the gate in accordance with a predetermined schedule. A better term for this control method is "adjustable flow-rate control." Such control method is very difficult to manage under unsteady flow conditions with frequent changes in demand and supply. Even under fairly stable conditions it requires excellent communication and good calibration of structures. In contrast, maintaining a constant water level at a check structure is relatively simple, even with manual operation, and does not require precise calibrations. Maintaining a constant water level at a check structure will result in constant flow rate through the turnouts located upstream. The flow through the check structure should be less important to the individual operator because he cannot influence the overall water supply to the system. For the system management at the head gate it would be sufficient to get feedback on flow conditions from the operator of the last structure or from any intermediate reach if the arriving flow is insufficient to maintain the desired water level.

Hydraulic Equipment

The main types of equipment used for regulation and control in irrigation systems are:

A. Equipment specific to canal systems

(1) Hand or motor-operated gates

- Overflow gates as cross-regulators.
  - Fixed composite (combination overflow and through-flow).
  - Adjustable composite structures (adjustable through-flow).
  - Various geometries (straight weirs compared with duck-billed weirs).
  - Various degrees of submergence.
  - Adjustable weirs.
  - Flashboards (wooden timbers used as an adjustable dam).
  - Drop-leaf gates
- Undershot (orifice-type) gates as cross regulators.
  - Multiple slide gates.
  - Single slide gates.

- * Radial gates.
- Power supply, electric controllers, motors.

(2) Automatic water level control

- Self-regulating gates
  - * Balanced gates operated by counterweights
  - * Balanced gates operated by controlled leak systems
  - * Float operated gates (AMIL, AVIS, AVIO, mixed types)
- Automatic motor operated gates
  - * Conventional motor operated gates as under (1).
  - * Gates with automatic controllers and feedback of water level, either electromechanical or with microprocessor.

(3) Flow rate control
- Proportional: flow dividers, fixed or adjustable.
- Manual:
  - * Manual gates plus d/s measuring flume
  - * Calibrated manual gate plus u/s water level control.
- Automatic:
  - * Modular distributor with or without u/s water level control,
  - * Motor operated gate plus microprocessor with feedback of water level over d/s weir or flume
  - * Motor operated gate plus microprocessor with feedback of u/s and d/s water level and gate opening.

(4) Measuring equipment
- Gate position sensors.
- Water level sensors.
- Flow measuring systems
  - * Flume or weir plus level sensor.
  - * Calibrated canal section plus ultrasonic velocimeter, level sensor and microprocessor.
  - * Calibrated gate plus level sensor, gate position sensor and microprocessor.
- Interfaces, analog-digital or digital-analog, for display, processing or transmission.

B. Equipment specific to pipe systems

(5) Isolating valves
- Hand and motor operated, mainly gate or butterfly valves.

(6) On-line flow or pressure control
- Manually operated valves (gate, butterfly, needle, multijet types).
- Automatic valves:
  - * motor operated valves as above plus microprocessors with feedback of pressure or flow rate.
  - * self regulating pressure reducers.

(7) Discharge or water level control at free surface head breakers or distribution structures

- Manually operated valves (as above plus self-centering disc and cylindrical types).
- Automatic:
  - * Motor operated valves as above plus microprocessors with feedback of pressure or flow.
  - * Float operated valves (cylindrical, self-centering disc, flap, various designs with hydraulically balanced shutters) for level control only.

(8) Discharge control at farm outlets

- Discharge limiting device.
- Gate valves, Californian valves, butterfly valves.
- Constant discharge valves.

(9) Safety equipment

- Pressure relief valves.
- Air exhaust and air inlet valves.
- Automatically closing valves:
  - * Butterfly valve combined with locked counterweight and velocity sensor acting on the lock.
  - * Buoyant plug-type valve to be installed over bottom outlet to vertical pipe.

(10) Measuring equipment

- Valve position sensor.
- Pressure sensor.
- Flow measurement devices:
  - * Differential pressure sensor plus diaphragm or venturi
  - * Electromagnetic, ultrasonic
  - * Calibrated control valves plus pressure sensor, valve position sensor and microprocessors.
  - * Volumetric meters, propeller meters.

C. Communication and data processing equipment

(11) Equipment for data processing and system management

- Microcomputers and industrial minicomputer.
- Peripherals (keyboards, VDUs, mimic boards, printers, mass memories, terminals)
- Software (system management, decision support, control, communication)

(12) Communication equipment

- Communication links between master, relay and remote stations via cable, fiberoptic, radio, microwave or infrared.
- Communication interfaces at remote stations.

D. Storage

(13) Reservoirs for volume control

- In-canal storage
- In-line reservoirs
- Off-canal reservoirs
- Service area reservoirs

The designer must fully understand the functions and the interaction of all equipment under different flow conditions.

Scheme Layout

The layout of the system will be largely determined by the relation of land and water resources, topography, and economic considerations. The question of intensive versus extensive irrigation has been discussed in Chapter 4. The design should incorporate, as much as possible, features that facilitate operation and provide flexible irrigation service:

- Use of low-pressure pipes instead of canals for tertiary distribution.
- Extra capacity in canals and pipelines (Gezira, Sudan).
- Buffer reservoirs and on-farm reservoirs (southern China, Friant-Kern Canal in California, Kano project in Nigeria, Jaiba and Manicoba in Brazil).
- Low level canals for privately owned low-lift pumping (Egypt and other delta regions).

The source of water may be uncontrolled or controlled. Simple run-of-the-river schemes with headworks that merely divert water into canals without controlling the diverted flow are called uncontrolled. These schemes are difficult to operate. Safety demands that such canals run low with ample freeboard, because the flow rate may change unexpectedly. With improved canal regulation, the same canals can be run at higher capacity since the emergency freeboard requirement can be reduced and planned operational spillage permitted.

Many projects are operated as "run-of-the-river" schemes, although there might be other alternatives. Frequently diversion schemes are located downstream of storage dams which can release water on an arranged schedule. The lag time between the dam and the end of the project can be partly or wholly compensated by:

- Storage in buffer reservoirs or large canals.
- Development of mathematical transfer functions that allow operators to better predict the downstream effects of various dam discharge (for example, Coteaux de Gascogne, France).
- Centralized control of cross-regulators to reduce the lag time of the system to the lag time of one canal reach (for example, the California Aqueduct).
- Appreciable operational spillage.

Schemes that are located downstream of hydroelectric power plants must accept water releases determined by power generation requirements rather than by irrigation requirements. Therefore, the design of the distribution system requires special considerations. If possible, afterbays should be built to make the irrigation system manageable (e.g Tadla, Morocco). If that is not feasible the preferable option is to provide some type of in-scheme storage. A good example is the so-called "melon-on-the-vine" design which is widely used in Southern China and consists of a large number of medium sized reservoirs and thousands of farm ponds.

One conclusion of this brief overview is that it is not enough that designers of irrigation schemes have long-term professional experience, what is also required is imagination and up-to-date skills. Technology is constantly developing; the electronic age has entered the irrigation sector. Increased client orientation, resource and economic constraints will require more efficient and responsive irrigation schemes. Engineers, like members of any other profession, must keep their skills sharp and have to undergo constant retraining and upgrading in order to provide designs that can meet present and future requirements.

6. CONCEPTS OF IRRIGATION DESIGN

Classical Approach to Irrigation Design

The classical approach to irrigation design tends to concentrate on various physical components such as dams, drop structures, canal lining, canal slope, and standard check structure without considering their interaction. The approach relies heavily on textbooks or manuals. The focus is on the structural details of the various components, proper analysis of stresses and moments, the types of nuts and bolts, reinforcement and so forth. Also important are the local availability and costs of materials and labor. The function and the performance of a particular structure can be fairly accurately determined if the construction is done according to the design. Such a design contains few unknowns, the task is well within the experience of most civil engineers, who are often not concerned with the overall "game plan" of the project.

In rehabilitation projects, traditional engineers might typically replace the canal banks with concrete lining without examining the need for a different flow-rate capacity or different regulating structures. They might also replace aging wooden structures with concrete and steel gates, without investigating the need and the possibility for operational changes.

The traditional approach to irrigation projects design seeks to maintain the status quo; it is not tremendously innovative nor is it much concerned with the eventual operation of the system. Mechanical and structural integrity analyses are emphasized. The approach to designing an irrigation project is similar to designing a building or a roadway. A steady-state flow is assumed for the purpose of designing canals and structures. This is not to say that proper structural integrity is not important. What is also needed, in addition to an examination of the statics and dynamics of the structures, is a detailed examination of how the structures perform jointly within the overall operational plan, especially where modern operations calls for variable flow conditions.

In some countries and institutions, there is considerable pressure on engineers to conform and to avoid innovative designs. Innovation, by its nature, involves risks. Many engineers avoid risks that may results in failures, entailing sanctions such as the transfer to some less attractive posts. Innovative designs require new construction techniques and new procedures, but most of all openness to new ideas and client orientation. This attitude must be supported from the highest ranks of the irrigation administration.

Evolution of the Design Thought Process

To produce a good final product, engineers must follow a certain evolution in their thoughts and in their perception of the design task. Table 6-1 describes the various steps in this evolution and the associated requirements. It is unusual for a person to simultaneously master all the steps. There are many irrigation designers who are fully competent in the first and second step, but relatively few who have gone beyond. Those who have progressed beyond the fifth step are no longer very competent in the first step, because they no longer deal with the details of structural design.

Table 6-1. Evolution of Irrigation Project Designer

Step	*Design Task*	*Requirements*
First	Classical structural design	Degree in engineering; access to design handbooks.
Second	Functionality of design	Some understanding of unsteady flow; ability to comprehend the function and purpose of various control devices including semiclosed pipelines; some field observation to verify the actual function of each component.
Third	Robustness, verification of equity	Exposure to a wide variety of structures; good understanding of flow in open channels; a deep appreciation for simplicity; understanding of the difference and importance of water-level control in some cases and flow-rate control in others.
Fourth	Ease of operation	Good understanding of and experience with operation of multiple structures with unsteady flow; extensive exposure to a variety of operational procedures and structures, however, a detailed knowledge of unsteady flow equations is not necessary; sympathy for harried operators and farmers.
Fifth	Consideration of water balance and quality issues in the hydrologic basin	Exposure to broad issues in water management; understanding of true definitions of irrigation efficiency at various levels within a hydrologic basin.
Final	Recognition of typical design errors	Experience and critical evaluations of one's own previous designs; experience with operation of irrigation projects, including impartial analysis of anticipated operation in contrast to actual operation and constraints.

It is bad practice if "standard or type designs" are inappropriately used or transferred from one scheme to another. Any good design needs a certain amount of site specific solutions. An experienced designer will consider explicitly all or most of the following aspects:

- Sizing of canals
- Functionality of control structures
- Robustness of design
- Ease of operation
- Water use efficiency
- Pipelines for final distribution
- Field application methods

Sizing of Canals

An important question should always be asked when designing a main canal system: Is it a design for a conveyance channel or for a canal that will distribute water under unsteady flow conditions? Both designs may have the same lining, slope, and cross-section. The difference between the two is that a conveyance canal just passes water from one end to the other with little need for regulation to control water levels. A canal designed for unsteady flow operation requires operational procedures and structures which will ensure reliability and responsive control for a variety of flow conditions.

A "water duty" or a certain flow per area is one way of determining canal capacity at the design stage as discussed in Section 4.5. In modern irrigation design it is not used for the sizing of canals since the operational target is to deliver certain volume of water at a flow rate that matches the needs of the clients. The question is, how can the proper size of an irrigation canal or pipeline be determined if the water duty is not used? Several aspects must be considered:

Operational storage

The size of a canal should not just consider the flow rates at some level of demand and delivery flexibility but should also provide some operational storage. Flows must be changed and routed from one offtake to another but changes of the flow rate do not occur simultaneously at all points along the canal reach. Operators cannot be everywhere at the same time, so the canal sections must often provide some buffer storage to temporarily store some volume between changes in deliveries. Of course, if the canal water levels are designed to fluctuate to provide operational storage, the offtakes must be designed to pass a nearly constant flow, even as the supply pool water level varies.

Shut-off capability

Canal cross sections should be large enough to provide some storage in case of unexpected closure of turnouts. An example is the Imperial Irrigation District in California. Lateral canals were originally not designed for much flexibility in frequency and flow rate. The cross sections became smaller from the inlet (large) to the tail end (small). Recently the district introduced a more flexible schedule. Farmers are free to shut off their deliveries at any time they desire. The changes of the operational procedure required that the tail ends of the canals be enlarged to handle the possible large flows. The cross section at the tail are now similar to the head end. Additional interceptor canals are provided at the tail ends to pick up the spill and divert it to buffer reservoirs for later use in canals at lower elevations.

Probability of demand

A flexible delivery system must be designed to provide water on days that farmers want to use the water. Such a system requires a larger canal capacity in the lower areas than needed for a strict rotation schedule, because several farmers may want to irrigate simultaneously. Several design strategies have been developed for sizing canals with varying levels of congestion. The formulas consider the number and size of the turnouts downstream in relation to the constant theoretical demand. In addition, these computations must consider the possibility that farmers may not irrigate at special hours or on weekends. The most widely used of these formulas was developed by Clement (1965).

Conclusion

The typical sizing procedure that accounts only for water duty is not sufficient for modern, flexible design. The design of canal and pipeline capacities at any point must consider:

- Turnout size.
- Number of turnouts downstream.

- Actual gross flow rate needed for continuous irrigation (that is, the water duty). This may be based on evapotranspiration rates or simply on some percentage of evapotranspiration which depends upon the degree of underirrigation built into the project.
- Probability of changing to crops with higher irrigation requirements.
- Probability of congestion (that is, the probability of not having enough capacity in the canal to provide the flow at a turnout when needed).
- Operational storage.
- Capacities required at the lower end of canals to handle flow-through of water that has been rejected at upstream turnouts, or installation of side weirs on the canal banks to discharge rejected water into drainage systems or recycling reservoirs.

A high density of turnouts is critical because it reduces the need for cooperation among farmer and it increases the command area by ensuring water delivery to all critical points.

Functionality of Control Structures

The first step in the professional evolution of an irrigation designer is to increase the understanding of how a particular structure interacts with other structures it its vicinity.

Location and interaction of structures

Three common problems relate to location and interaction of structures:

- Many conventional canal designs have too few check structures to ensure adequate water-level control at outlets. Thus, the flow rate through the outlets varies with time even though it is supposed to remain constant.

- The difference in head across structures is frequently insufficient. Larger changes in elevation will dramatically stabilize the flow rate at standard underflow turnouts. Drops across farm turnouts recommended in some design manuals are insufficient.

- The designs of lower-level (tertiary, distributary, and watercourse) distribution systems are frequently not based upon sufficient topographical data, with the result that the supply system may be at a lower elevation than the destination of the water.

Orifice compared with weirs for turnouts

Turnouts (offtakes) at all levels in modern irrigation systems can have three functions: (1) on-off, (2) flow-rate adjustment and control, that is, to maintain a constant, desired flow rate through the offtake, and (3) flow-rate measurement. Many projects have offtakes suitable for one or two of the functions, but inadequate for the others. Recommended simple designs include modular distributors, which serve all three functions and single undershot gates (orifice design), which can be configured to perform all three functions, if enough head is available.

Offtake designs that are not recommended are those with weirs, such as Rominj gates (mobile adjustable weirs widely used in Indonesia), especially if they are associated with undershot cross-regulators in the supply canal. In this case, a slight change in the water level of the supply canal will result in a large percentage change in flow through the weir offtake (see Section 7.5.2).

Orifice compared with weirs for check structures

Manually operated check structures (cross-regulators) should include the hydraulic advantage of a weir in part or all of the structure. The weir will minimize the upstream water-level fluctuation even if there is a moderate change in flow across the structure.

The concept of using weirs in the supply canal and orifices at the offtakes in manually operation of canals is very simple and does not require much understanding of complex hydraulics. Nevertheless, there are many projects designed with exactly the opposite configuration.

Automatic water level control

In many cases (especially on small canals), proper application of hydraulic control principles and float gates can provide relatively inexpensive and sufficiently accurate control of water levels. In larger canals electrical, microprocessor-controlled gates are used which have the advantages of providing precise water-level control, the ability of changing the set points, and the possibility of functioning in response to remote commands. There are, however, some disadvantages with respect to costs, reliability and maintenance. The robustness of hydraulic gates sometimes results in better overall water levels control than with electronic equipment.

Robustness of Design

Robustness of design denotes the capacity of structures or equipment to perform under adverse conditions. Robustness is a very important feature under real world conditions, although it is difficult to define. Designers with actual operational experience are more inclined to produce robust designs.

Specific control structures that rarely fail

Examples of good robust control structures include the following:

- **Long-crested weirs** (duck-billed or oblique weirs). These structures are extremely simple and durable, in particular, if gates are installed at the downstream end to flush out silt. In Portugal, new irrigation projects are being equipped with long crested weirs, even though automatic hydraulic gates were used in the past for the same function.

- **Manual radial gates with side weirs.** The radial gates are used for large flow adjustments whereas the side weirs, generally with a shorter crest than long-crested weirs, compensate normal fluctuations.

- **Canal offtakes and metering gates.** Canal offtakes and metering gates are frequently used in old projects in the western United States. They serve as on-off, flow-measurement, and flow-adjustment device. They control the opening to a short pipe section between the canal and the field; the gate opening and the loss in head across the gate provide the data necessary to determine the flow rate from a table. They can supply either ditches or pipelines. They are not to be confused with the constant head orifice offtakes which have two gates per structure. Their accuracy is not superb (+/- 7 percent) but they are very robust.

- **Modular distributors.** These also serve as on-off, flow-measurement, and flow-regulation devices. They serve as offtakes from one canal to another canal, or from a canal to a field. Additional advantages include easy flow measurement, insensitivity to upstream water level changes and the ability to lock them at a specific flow.

Specific structures to avoid

Some poorly chosen devices are still found in new projects and often contribute to the impression that water-control structures are ineffective. Examples include:

- **Rominj gates.** The weir action creates large fluctuations of the flow rate in offtake with minor changes in the water levels of the supply canal. The World Bank sector review on Indonesia (ASTAG, 1990) observed:

  ".. current design guidelines have favored the adoption of Rominj gates because they provide both control and measurement capability in a single structure. Very often the solution adopted has been sluice-gated controls along the parent channel combined with Rominj-gated offtakes and outlets. Unfortunately this is possibly the worst of all combinations from the hydraulic point of view since it is extremely unstable and requires high construction standards, accurate calibration, and disciplined operation if correct discharges are to be maintained. Small deviations have a proportionally great effect. The reason why this system has persisted lies largely in the excellent metering potential of the Rominj gate when considered in isolation. The negative effect on system operational stability has been overlooked."

- **Vertical slide gates as check structures.** Slide gates have three disadvantages: (a) they are underflow gates, therefore the upstream water level tends to fluctuate widely with small adjustments; (b) they require considerable force for operation; (c) they tend to stick in place easily.

- **Radial gates as check structures without side weirs.** This option should be avoided in manually operated systems. Although radial gates have operational advantages (less force, little blockage), but they are basically underflow weirs with the same problematic hydraulic features as slide gates.

- **Constant-head orifice (CHO) offtakes.** This type of offtake structure can be found in many schemes throughout the world. The basic hydraulic assumption is a stable water level in the supply canal or relatively insignificant fluctuation between adjustments. Since this is frequently not the case, the adjustments on the offtakes must be estimated. As may be observed in virtually all projects in developing countries, the staff rarely use the CHO gates for water measurements because of the relative complexity of operation (Plusquellec, 1989).

Ease of Operation

There is considerable debate among irrigation engineers regarding the desirability of having adjustable structures. Many writers have expressed their preference for proportional, nonadjustable flow dividers on the grounds that the operation is simple, understandable, and ensures an equitable division of water (Horst, 1987). Some structures make it particular easy to verify the equal distribution of water through simple visual observation. Anybody, just by looking at these structures, immediately knows the flow rate or division of the flow rate at the bifurcation point. Devices with special significance are modular distributors and adjustable proportional division weirs.

Simplification is desirable and often achievable but should not come at the expense of the quality of irrigation service. There is substantial evidence that reliability and equity are not necessarily the consequence of using simple structures. Proper selection of hydraulic equipment will often greatly simplify operation and maintenance while providing better irrigation service.

Modern design concepts and equipment can assist the designer to:

- Reduce the amount of information that needs to be generated and transferred within the system.

- Reduce the number of manual gate adjustments by using appropriate combinations of fixed and automated structures.

- Create a design that is not affected by imperfectly shaped canals. There is a high probability that unlined canal sections will change due to erosion or siltation. Lined sections can also change due to siltation, vegetation or accretions of small shells. Appropriate designs should provide for structures that are simple to operate and maintain.

- Eliminate the need for any complicated set of daily instructions and detailed real-time modeling as a requirement for daily operation.

Water Use Efficiency

On-farm irrigation and project efficiencies can be meaningless when describing total water use efficiency. Often water is recoverable lower down in the system, but is lost to the originating area. Assessment of the total water use efficiency requires that the water balance of the whole hydrologic basin is considered. Two items are particularly important in this regard:

- **Unlined canals.** If unlined or leaky canals recharge reusable groundwater, it may be better to concentrate on improving irrigation services rather than spending time and funds on canal lining.

- **Spill.** Spills may reduce the project's delivery efficiency, but may not be a loss to the hydrologic basin. Operational spills are quite acceptable if they enable the operators to achieve a high degree of flexibility in water deliveries. If the system is designed to handle spill without adverse consequences, a moderate amount of spill may actually improve overall project efficiency by increasing on-farm irrigation efficiency. Examples of appropriate use of spill are the Coello project in Colombia and the Palo Verde Irrigation District in California. This argument should not be used, however, as a "catch-all" excuse pretending that water control is not important.

Pipelines for Final Distribution

Perhaps the single most neglected option for irrigation design is the pipeline. Pipelines may be closed, open, or semi-closed, depending upon the situation. Open and semi-closed pipelines with float valves, permitting the use of non-reinforced concrete pipes which can be very economical in diameters of 200 mm and larger. Pipelines have tremendous operational advantages over canals at the lower levels of a distribution system. Some of the advantages are the following:

- **Simplicity of operation.** Depending on the particular design, pipelines can provide water without the lag time. Pipelines are ideal for day-only irrigation, since they remain full of water during the night and are ready for instantaneous startup the next morning.

- **No Spills.** Closed and semi-closed pipelines have the ability to transmit pressure, They automatically match the inflow with the outflow, which can be regulated at the delivery point. Thus semi-closed systems on steep ground and open systems on flat ground can be operated without any spill.

- **Reduced maintenance.** Pipelines, because of their flow velocity and turbulence, rarely retain silt. Weeds do not grow in pipelines. Farm equipment and cattle do not damage pipelines.

- **Less land out of production.** No obstacle to farm equipment. Right-of-way is more easily obtained since alignment is less restricted. This is particularly important with small fields, where the distribution canals may occupy a significant amount of land.

- **Increase of the command area**, due to the ability to serve areas which are out of gravity command. Low lift pumps can supply low pressure pipelines from canals and reservoirs.

Unfortunately many new pipeline systems are sized only on the water duty and are mainly installed because of no seepage and no spill advantages. Small increases in the size of the pipes will produce major operational benefits due to increased flexibility of operations (Merriam 1992).

Field Application Methods

The success of a project may well depend on the design and selection of the appropriate on-farm irrigation method. This is especially true where more modern techniques such as gated pipes, drip, or sprinkler irrigation are used. But also improved surface irrigation methods (furrow, border-strips, basins) are sensitive to variations of stream flows and duration. It is also essential to understand that the on-farm irrigation methods have specific water-delivery requirements. Consider two examples of inappropriate use of modern on-farm irrigation technology:

In the Doukkala project in Morocco, it was assumed that farmers would share sprinkler equipment and adopt a specific crop-rotation pattern. The choice of sprinklers for on-farm irrigation was correct because of the topography and soil conditions. The assumption that farmers would share the equipment was incorrect. Each farmer wanted his own sprinkler line in his small field, which did not conform to the optimum spacing of the selected sprinkler equipment. As a result, the on-farm irrigation efficiency is considerable less than expected (OED, 1989).

Many on-farm irrigation improvement programs, supported by external funding agencies, envisaged the use of modern technology such as infrared thermometers, neutron probes, complicated measurements of hydraulic conductivity, etc. as means to improve irrigation efficiency. Such technologies are rarely used in production agriculture, even in areas with advanced irrigation systems and a long history of irrigation improvements (e.g., California, Israel, Australia, France). Simple steel probes, as mentioned in chapter 4.5.3, are more practical under most circumstances.

Judging from experience the authors contend that the performance of on-farm irrigation critically depends on quality of irrigation service of the main system. Table 6-2 describes the relationships between on-farm water use efficiency and some delivery factors.

Table 6-2. Potential Water Use Efficiency with various on-farm irrigation methods as affected by water deliveries

	With good water level control		*Without good water level control*	
On-farm irrigation method	*Arranged*	*Rotation*	*Arranged*	*Rotation*
Sprinkler	high	not feasible	high	not feasible
Upland crop-surface irrigation	mod-high	mod-low	mod-low	low
Rice-large turnout flow	high	mod-low	moderate	low
Rice-small turnout flow	moderate	low	low	low
Drip	high	not feasible	high	not feasible

Note: "Good water level control" refers to a constant canal water level at the turnout, thereby assuring a constant flow rate through the turnout. No under-irrigation and no on-farm storage is assumed.

On-farm irrigation designs are as diverse in their performance as are various water-delivery designs. The outcome depends not only on the quality of design but also on the management input. Figure 2 illustrates some general relationships. It should be noted that modern on-farm irrigation techniques such as drip irrigation perform well only with good management and water-delivery service.

Figure 2 Relative efficiency of on-farm irrigation methods in relation to management input and design

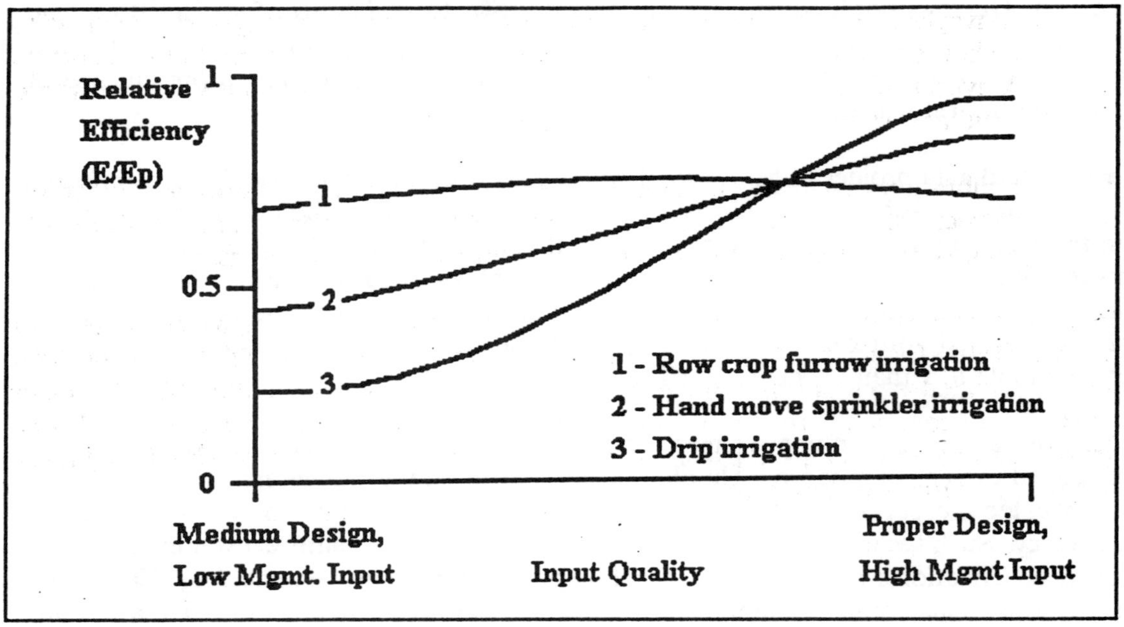

Typical Design Errors

Extreme simplicity in design and operation has often serious drawbacks. In particular, many simple and nonadjustable structures will not function well if the conditions change on which the design was based. Many extremely simple systems are so rigid that the operational procedures cannot be changed when required. Therefore, design of irrigation systems must allow for the possibility that the design assumptions are not correct or change during the life of the project. The design must include sufficient flexibility to adjust the operation procedure. Murray-Rust and Vermillion (1989) observed the following in Indonesian irrigation projects:

- In many systems in east, central, and west Java, even after many years of operation, data generated during the design process is used without updating information on the system status.

- In all the systems studied, there have been substantial changes over time in area, cropping patterns, and system status. This means that assumed design and operation values are no longer valid.

- Estimation of water availability for irrigation. There were often gross hydrological errors, resulting in over- or under-estimation of water availability, both during the year and various seasons.

- Estimation of field level demand for irrigation water. Design assumptions of field efficiency, cropped acreage, and cropping patterns were frequently incorrect.

- Conveyance losses in main and secondary canals. Errors in the estimate of conveyance losses resulted in water levels throughout the system which were higher or lower than those predicted during design. When these incorrect conveyance losses were used in sizing and locating unadjustable structures these structure were found useless in actual operations.

- Calibration of gates and measuring devices. Simple delivery systems depend on correct calibration of structures for proper hydraulic performance. There are serious difficulties in implementing the correct field calibrations.

In another case study of the Tungabhadra project in Karnataka, India, Murray-Rust and Snellen (1991) found that "...although the design was based on a specific command area and an assumed value of losses along the canal, post-construction addition of outlets and expansion of command area has meant that the total sanctioned water supply and losses greatly exceed the design capacity of the main canal....This type of situation makes operational targets meaningless."

7. DESIGN ASPECTS OF SUSTAINABILITY

The sustainable operation of irrigation projects involves a physical dimension (maintenance and prevention of environmental damage), a financial dimension (cost recovery), and a social dimension (user participation and ownership). Some aspects of all three are discussed in this chapter to the extent that they relate to the need for improved water control.

Maintenance of Modernized Schemes

Two major concerns preoccupy irrigation managers when they consider modernizing their scheme with automatic regulation structure and electronic communication equipment: removal of silt and repair of mechanical and electronic equipment.

Before the issues are discussed in detail, some points need to be made. Irrigation agencies and governments have to make serious efforts to address the maintenance problem. Designing for substandard performance to match substandard operation and maintenance practices cannot be an option for the future. In most projects there is an urgent need to establish realistic and properly funded maintenance plans (ICID 1989). Water rates have to be collected and reinvested in the irrigation schemes (OED 1993). Only if these basic preconditions are in place does it become possible to successfully address the maintenance problems.

Removal of silt

There is an inherent conflict between flexible delivery and maintenance costs in run-of-the- river schemes with high sediment load. Flexible delivery results in unsteady flow conditions and occasionally low flow velocities, thus increasing siltation. For any system with high sediment load and some flexibility, there seem to be three options:

- remove the silt before it gets into the canals;
- make it easy to remove the silt from the canals;
- keep the silt in suspension.

Some design modifications may help in reducing the amount of silt that settles in the distribution system. Large desilting basins can be provided at the canal heading. The entrance to the All American Canal from the Colorado River is a good example of such a desilting basin. Other less costly alternatives include vortex tubes and tunnel extractors. The intake of the canal can be designed so that only water from the surface is diverted (skimming). The Coello Irrigation District in Colombia (Plusquellec, 1989) uses such a design. In the canal itself, long-crested weirs can be used with the nose pointed downstream. Silt will accumulate behind the weir. A gate at the nose can be opened to flush the silt out; if possible the silt should be diverted back into the river. Observations in California have shown that long-crested weirs that are silted up to the crest can be flushed out within a few hours.

Silt removal from within canals themselves can be simplified by making bottoms wide enough for mechanical clearing and providing good access along the canals. Lined canals with steep slopes can maintain sufficient velocities to prevent sedimentation (for example, the Yellow River basin in China).

Maintenance of mechanical and electronic equipment

Mechanical and electronic equipment always requires some form of regular maintenance. The following aspects of this maintenance must be considered during design:

- **Frequency of maintenance.** Some structures, such as float-operated gates, only require lubrication once a year. Electronic controllers require more frequent preventive maintenance.

- **Quality of maintenance.** Electronic controllers require specialized equipment and staff who must be familiar with the particular device. Long-crested weirs and float-operated gates require only the "pick and shovel" qualification of maintenance staff.

- **Anti corrosive protection.** Painting can be minimized by specifying aluminum or fiberglass gates. Corrosion of metal parts can be reduced or eliminated through the use of cathodic protection or galvanization.

- **Replacement parts.** There is a fundamental rule with respect to electronic control equipment in irrigation projects: avoid making your own electronics equipment, even though it may look inexpensive. Canals are not research projects or classrooms. They must work continuously, and if automation equipment breaks down, the parts must be replaceable immediately. This requires off-the-shelf components of modular design. Everything must be commercially available. Projects that try to develop their own electronic equipment typically suffer from three problems: (1) the equipment rarely works, (2) when it finally works, production is behind schedule and the costs are many times higher than anticipated, and (3) the equipment breaks down more frequently than commercially available goods.

A major problem in many irrigation projects is vandalism. The large area covered and the remoteness of many locations makes it difficult to protect valuable equipment. A solution is difficult. Most promising is an approach combining sanctions, incentives, creation of a sense of ownership, and education. Sensitive equipment should first be installed in places that offer some kind of protection.

Cost Comparison and Cost Recovery

Economic analysis is an essential part of project preparations. Unfortunately, standard procedures of economic analysis, especially the overriding importance assigned to the economic rate of return (ERR), favor low-cost equipment and structures because of the discounting of future benefits. If only the direct cost of traditional and modern irrigation equipment are compared, modern concepts may not be considered because of higher initial costs. The advantages of modern design concepts become evident if the increased benefits in terms of higher water use efficiency, increase of irrigated area, increase in quantity and quality of crop production, reduced and more convenient farm labor are taken into account.

A realistic comparison must consider the following:

- Total project costs include costs of planning, land acquisition and compensation, construction of roads and utilities, water diversion or storage, resettlement, distribution system, pumping stations and drainage. A 20 percent increase in the cost of a canal or a pipeline may only represent a 5 to 10 percent increase in the total cost of a project. The substantial increase in water use efficiency and project performance might well justify the additional expenditures.

- The cost of improved structures and control systems is often relatively minor when compared with standard improvements, such as canal lining. Furthermore, sizing canals for increased capacity is also a relatively minor cost when compared with the total construction cost of a canal. Doubling the capacity of a pipeline increases the costs of the pipe by less than 20 percent and the project costs by only about 5 %, but provides much-improved irrigation service.

- Existing irrigation projects tend to have overall efficiencies in the range of 25 to 35 percent. This means that 65 to 75 percent of the water is not being beneficially used on the project. An economic justification of improved water control must include an estimate of improved efficiency.

- Increased crop yields are perhaps the biggest benefit derived from better water deliveries. For example, in California the cost of field irrigation improvements is more than compensated by increased yields or better crop quality rather than by water savings. The same benefit occur in developing countries to an even greater extend because the improvements are of greater magnitude.

The potential for cost recovery must be realistically considered. If farmers are to be charged for irrigation, but do not receive reliable and equitable service, it is impossible to collect water fees. Without improvements in the level of service collection of water fees to pay off project loans is unrealistic. In a related statement, Bottrall (1979) remarked:

> "...despite the desirability of raising water charges, it should be seen as a secondary issue in terms of sequential actions, first because it is a highly politicized issue; and secondly because in most cases farmers will not become better disposed to the idea of higher charges unless other changes are made first--the most important of which is an improved water distribution service."

Bottrall continues:

> "...wherever water is scarce in relation to farmers' demand for it, the most important of all functions of irrigation project is to ensure that it is efficiently and fairly distributed. It is only if the main water distribution system is well operated that many other important management objectives can be satisfactorily realized (such as "on-farm development" work, improved watercourse and farm-level water management, the introduction of higher water charges); and it is only then that high returns can be

obtained from agricultural extension advice and the increased application of other complementary inputs...."

Although made 15 years ago, this statement describes the actual situation in many schemes well. In conclusion, the comparison of the total cost of many modernization projects with the present levels of performance and potential improvements indicate that there is sufficient economic justification for investment in improved water control systems.

Participation of Water Users

Most new or modernized irrigation projects include some kind of water user association (WUA) in their organizational structure. Many studies (Freeman, 1985, Uphoff, 1986) underline the need for a local intermediate organization between the irrigation agency and the individual farmer. The following section discusses the functions of WUAs with respect to the water-delivery systems.

Functions of water user associations

WUAs have important functions in water distribution, fee collection, maintenance, conflict resolution, and representation of farmers in discussions with public agencies. In order to fulfill these functions, WUAs must be legal entities that can enter into contracts, open and operate bank accounts, and sue and be sued. They must have the power to enforce rules and regulations and control freeriders. The legal system must protect the association's right against irrigation agencies.

The viability of WUAs, and their ability to perform their functions, depends in large part on the reliability of water delivery to individual farmers which in turn depends on reliable supply from level to level in the system. The interaction between agency and WUA is of particular importance. Bulk supply, measurement devices and buffer reservoirs which contribute to reliable water delivery greatly support the functioning of WUAs.

Realistically, the immense length of tertiary distribution and field canals or pipelines can be maintained only through WUAs. WUAs are, thus, extremely important for the long-term sustainability of irrigation projects.

Water management functions

It is generally assumed that the water agency will deliver water in bulk to a certain point in the system. At that point, water distribution becomes the responsibility of the WUA. In most cases designers have only vague ideas about how the water distribution within the WUA area will actually carried out. In fact, it is often assumed that within a WUA the farmers themselves will somehow personally perform the water delivery. The argument that control structures must be very simple and easy to understand is often based on the belief that farmers will actually operate the system.

Farmers must understand the structures and operational procedures if they are expected to assume responsibility. However, farmers in well managed WUAs do not operate the distribution gates themselves. They may operate their own turnout in a few cases, but nothing more. Well managed WUAs hire professional staff for the actual water management within their area (Plusquellec, 1991). The members of the WUAs only agree on the rules and regulations to be followed by the hired employees. This procedure reduces potential conflicts among individual farmers. Examples of WUAs that hire staff to manage water deliveries include the Coello district in Colombia, farmer managed irrigation districts in Mexico, village irrigation teams in China, some water user associations in India and all the USA. This approach is vastly different from direct government control where government employees control all levels of the irrigation system including the farm level.

Modern technology and water user participation

Frequently the argument is put forward that automatic water control will limit farmers' participation and that, therefore, manual control systems should be preferred. The assumption is that farmer participation in the operation of larger canals is desirable. In schemes organized according to the service concept, however, it is not necessary that farmers or farmer representatives interfere with the operation of the main system. In successful modern irrigation projects the details of maintenance, automation, and operation of the larger canals are not dealt with by the water user associations; those are clearly agency operations in most projects.

The authors of this paper do not argue for complicated control devices at the lower levels of distribution in a project. In most cases, proper hydraulic design and simple automation techniques will make significant improvements. Electrically operated gates, computers and microprocessors may be required sometimes, but their use is generally limited to larger canals. A good design, even with complicated devices, will result in simple rules for the actual operation at all levels in the system.

Reliability of supply and cooperation among farmers

Regardless of the type of the control structures, successful WUAs require significant farmer involvement in decisions making on rules of water delivery, settling of disputes, collection of funds, financing of improvements, and maintenance. It is important to design a distribution system that prevents conflicts over water distribution and inflexibility of supply from dominating the business of water user associations.

Reliable water delivery is a key to achieve farmer cooperation. Reliability is not to be confused with adequacy. Reliability defines the certainty with which water will arrive at the promised time, in the promised amount, and for the promised duration. It also considers the magnitude of flow-rate fluctuations. As reliability decreases, the potential for conflicts increases. Uncertain delivery results in suboptimal utilization of labor and farm equipment. With very unreliable deliveries, yields begin to decline and cooperation declines. At that point, farmers begin to look seriously for ways to minimize their individual loss.

Historically it has been difficult to form a viable water user organization with extensive irrigation at a subsistence farming level. In Asia many farmers operate on a subsistence level on very small fragmented holdings and are receiving only extensive irrigation. WUAs are difficult to sustain under these conditions. On the other hand, there are many examples of functioning WUAs in traditional irrigation schemes with small holdings and

a scarcity of water. The reason for the difference is probably the difference in governance; that is, the social and judicial system that enables smallholders and WUAs to assert their rights.

8. THE DEBATE ON MODERNIZATION

The final chapter is intended to contribute to the debate on new approaches to irrigation design. First, the reasons for the slow process of transfer and adaptation of modern water control concepts are examined. Then, some arguments against modernization are discussed. Third, a summary of the report with a strong plea for a new approach to irrigation design is presented.

Reasons for the Slow Transfer of New Concepts

Despite the many advantages of new concepts as discussed in the previous chapters, the fact is that the adaptation and transfer process of modern technology in irrigation has been slow. Adverse administrative and behavioral reasons include:

- Pressure from lending organizations to shorten the implementation period. Project implementation will be delayed if traditional design criteria and standards are questioned or new concepts tested in pilot schemes.

- The lack of economic pressure on irrigation agencies and contractual motivation for their consultants to introduce new concepts.

- Resistance to change by irrigation managers, engineers, and others. Risk aversion and adherence to outdated design manuals.

- Lack of operational experience and service orientation by planners and irrigation departments. Top-down design and management more often reflects a state of mind than a technical necessity.

The fundamental cause for the slow rate of technology transfer, however, is a lack of knowledge of available technologies and a misunderstanding of the nature of irrigation, in particular :

- A lack of sufficient training at all levels, from the university to the field. Training needs include agronomic and engineering principles coupled with specific technical details for designing and operating modern water-delivery control.

- Misdirected pride of individuals, agencies, and countries. This pride results in an unwillingness to search, examine, and adopt new ideas developed elsewhere. It also prevents individuals and institutions from objectively analyzing existing local conditions and performances.

- A lack of understanding of on-farm needs by planners, designers and administrators.

- A severe lack of well-documented studies on the performance of modern irrigation schemes, the noteworthy exceptions being J. Merriam (1990, 1991, 1992).

Discussion of Arguments against Modernization

There is hope that the above obstacles can be gradually removed through changed administrative procedures, appropriate incentive, improved training, study tours, increased dissemination, and publication. However, some writers have put forward arguments in support of the view that modern water-control concepts are generally misplaced in irrigation schemes of developing countries. The most frequently heard arguments in the debate are:

- There is little need for flexible water delivery, the agricultural benefits are relatively minor.
- Water is a scarce national resource that must be controlled to secure equal distribution over a wide area.
- Water in run-of-the-river schemes cannot be saved and should be used for groundwater recharge as much as possible.
- Management skills are the most constraining factors in developing countries and therefore irrigation schemes should be as simple as possible.
- Modern technology is difficult to maintain and vandalism cannot be controlled.
- Modern technologies and design concepts are too costly.

These arguments need to be addressed in more detail.

Suitability of modern technology

Burns (1993) discusses appropriate water control technologies for monsoon areas in Asia. He supports provision of intermediate and on-farm storage, maximum utilization of privately owned, low lift pumps and the service concept in general. He forcefully argues against attempts to solve the many problems of large gravity irrigation schemes through improved canal regulation, adjustable gates, and arranged demand operation, because of the complexity of irrigation in areas with highly variable rainfall and the inherent rent-seeking behavior of farmers and bureaucrats alike. His recommendations are: (1) give up on the myth of just-on-time operation; (2) adopt the structured approach for design and operation; (3) consider water as a utility for which the privileged have to pay; (4) monitor performance; and (5) consider the use of lift irrigation technology and interim storage.

It should be clear that the authors of this publication do not argue for manually operated, gated systems nor for on-demand operation in extended gravity systems. However, modern technology and appropriate automation have the capacity to solve many of the managerial problems of large systems. Although some difference of opinion remains, the authors agree with most of the other recommendations under conditions described. Several examples of schemes with buffer reservoirs and individual pumps are presented in Part II.

Benefits of flexible water delivery

People contend that there is no need for flexible water delivery. They say that most crops in developing countries grow well with regular rotational supply and that farmers can easily adjust their cropping pattern and farming practices to predetermined irrigation schedules as long as these are reliable (Perry, 1992). According to them, the additional benefits of flexible supply are not sufficient to balance higher investment costs and operational complications.

The argument is valid for flat areas with deep, homogenous soils, uniform cropping pattern, standard crops, and well distributed rainfall such as in the northern Indian plain. Many other areas, however, have wide variations in topography, soils, crops and rainfall. As discussed in Chapter 3, farmers in these areas would greatly profit from flexible water supplies. In most irrigation schemes the cropping pattern is diversifying rapidly and farmers, even in the favored plains, are prepared to pay very high premiums for flexible water delivery from private wells. Farmers in the United States are constantly modernizing their irrigation equipment and irrigation practices in an attempt to obtain higher yields and to save labor, not to save water. Higher yields depend on flexible water delivery, either from the main system or from on-farm storage.

Management of scarce water resources

The design logic of extensive irrigation, water spreading and central administration, developed by the colonial powers in India and Pakistan, has served the countries well as long as rules were enforced, holdings were sufficiently large and overall demand for food was smaller. It has now being observed that the central administration of scarce resources often leads to rent seeking and excessive political influence. It has also been observed that spreading water too thinly increases the costs of production and does not generate sufficient income at the farm level to cover operation and maintenance costs. Some of these problems can be solved if water is treated as an economic good. The notion has gained acceptance since the Dublin Declaration and is reflected in recent water resources policy papers (World Bank, 1993). Implementation of the notion demands wide ranging application of modern technologies for improved water management.

Lack of storage in run-of-the-river schemes

Irrigation schemes that are supplied through river diversions without internal storage the only and most valuable storage may be the aquifer. In such cases there is indeed little need for precise flow and water-level control in the main system except for safety purposes. As much water as possible should be diverted during the flood season into the irrigation area to recharge the groundwater for later use in the dry season. This management concept has been very much refined in China, where people talk about managing the four waters (Feng and Wolter, 1993). Modern water control concepts are most valuable in schemes that include upstream reservoirs or substantial buffer storage.

Management skills as a constraining factor

Horst (1991) is a prominent advocate of the argument that management skills are the most constraining factor. He argues that irrigation technology with adjustable gates, which worked well under conditions of strong governance, cannot be sustained because the management and law enforcement capacity of many developing countries is inadequate. The suggestion is that the technology must be simplified and that the so-called structured approach should be introduced, meaning that irrigation systems should be managed only to a certain level (distributary level), below which the canals run in an on-off mode and water is distributed through fixed outlets and proportional dividers.

When weighing the argument it may be useful to consider the findings of a recent study on a medium sized project in Mexico, that produced some insights in the management capacity of local organizations (van der Zaag, 1992). The scheme is entirely equipped with adjustable, under-shot gates. Under-shot gates are known for their low operational performance, they make it difficult to stabilize flows and water levels in the system. However, the "canaleros" have managed to make the scheme work and to provide good irrigation services despite the inherent technical deficiencies of the infrastructure. Adequate financial incentives, technical competence and dedication have contributed to overcome technical problems. Only recently has the National Water Commission of Mexico decided to modernize the control system (see Part II: Mexico).

An in-depth debate on the relative importance of management skills would exceed the scope of this paper. Even if it is the most important constraining factor, there are other solutions than simplification of irrigation through the structured approach. With good design, operations can become easier and more reliable. If assisted by suitable technology operators will be able to manage water rather than to struggle constantly to avoid spills and other disasters. Actual operation of the structures should be done be hired professionals, not by farmers. This practice should reduce the shortage of management capacity.

Cost of modern irrigation technology

This issue is discussed in Chapter 7. It is contended that the additional costs of better water control technology are offset by increased yields and ease of operations. Preliminary cost comparisons of two modernized schemes in Indonesia and Mexico (see Part II) led to the conclusion that modern control systems do not substantially increase the investment costs if considered early in the design process. Further detailed case studies and project monitoring are required to substantiate this hypothesis.

Maintenance and prevention of vandalism will remain a problem as long as the local population does not relate the new technology to improved services and increased production. There is no substitute for good governance and social control. Repair of equipment will gradually become easier with the development of the service sector and modular design.

The Case for Modernization

The authors believe that the case for modernization of irrigation systems is compelling. Low project efficiencies, dismal performance, negative environmental consequences, premature failures or abandonment, and inequitable and unreliable water deliveries are well documented. On the other hand, increased world population, lack of new land, and environmental awareness demand that irrigation projects be better-managed and more productive. The authors contend that a proper blend of education, awareness, and initiative are the ingredients needed to promote successful modernization of irrigation systems. New design concepts, design tools, equipment and communication systems exist; progress requires that designers and planners become familiar with such tools and concepts and apply them.

Modern design is a thought process that starts with the definition of the operational plan. A good modern design makes full use of advanced concepts in hydraulic engineering, agronomic science, irrigation engineering, economics, and social science to identify the simplest components and a workable solution. The design process begins with the identification of performance objectives and the definition of the operation plan. Modern design must not be confused with new physical components in a system. Some modern designs use very simple water-control devices; others may require sophisticated controllers and communications equipment to achieve a desired level of performance. These component should be selected as a last step in a design process.

The operation plan should be based on a comprehensive or holistic view of the project. Some of the important factors to consider include the nature of the crop, soil type, water availability, silt load of the water, climate, technical infrastructure in the region, farmer skills and cooperation, value of labor, and farm size.

Successful modernization requires an understanding at all levels (planners, designers, construction specialists) of the broad objectives and performance standards of irrigation projects. The objectives and constraints at the farm level should be of overriding concern because it is the farmer who has to generate the surplus to make the project sustainable. Designers must understand that the farmers can increase production only with reliable and convenient water supply. In turn they will pay more, if labor for water application is reduced and made more convenient, for example through the reduction of night irrigation.

The authors strongly argue for a "bottom-up" approach to design with maximum participation by users. There are a wide variety of designs concepts, structures, methods of control, and schedules which may lead to many different combinations. It is essentially the user, who has to be satisfied with the quality of service at the downstream ends of the system. Design engineers must be aware of the resource limitations and of the implications of their design for maintenance, operation, and flexibility of water use.

Continuous training of design engineers is crucial to modernization. Training must be provided in (1) the need to provide convenient service to clients, (2) modern automation equipment, and (3) strategies of water control. Classical structural design (concrete lining and support structures) is very important, but it is only the basis of a good design. Designers must have a clear vision of how their product will be operated in the field. They must develop clear operational procedures and should be able to follow their own instructions in the field, once water begins to flow.

In summary, installation of new equipment can improve performance only if operators and designers have a common and well-defined vision of operation procedures and maintenance requirements, if performance standards are precisely defined at each level, and if there is an appropriate incentive structure.

PART II: NOTES ON SPECIFIC PROJECTS

Examples of good, innovative irrigation design that could illustrate aspects discussed in Part I exist throughout the world and provide lessons for designers and managers. This part of the paper presents about two dozen projects of fifteen countries to illustrate particularly interesting technical solutions of operation and management problems. A summary of these projects is provided and common features are discussed followed by a brief description of technical solutions of the individual projects. The descriptions are based on field notes from relatively brief visits and on material supplied by the respective project agencies. No systematic detailed performance analysis or design studies have been done by the authors. Omissions and misunderstandings are therefore possible.

9. MAIN FEATURES

The selected projects represent a wide range of agroclimatic and socioeconomic conditions, from the arid zone of northwest Mexico, California, central Sudan, the Nile Valley, central Iraq, the moderate rainfall zone of northern Nigeria, the coastal plain along the shore of the Caspian Sea in Iran, to the humid tropics in Malaysia and Indonesia. Average farm size in all these projects is small (less than 5 ha) with the exception of the three projects in the United States and the Jaiba project in Brazil.

In all the projects individual farmers enjoy some degree of flexibility in water delivery. In most cases, irrigation scheduling is arranged between farmers and irrigation agencies. Central scheduling is only used in Egypt, Sri Lanka, and Malaysia, but Egyptian farmers can use low-lift pumps at any time when their canal is "on." In the Muda project in Malaysia, the recent introduction of low-lift pumps provides better control over timing and the amount of irrigation water as required for mechanized farming and direct rice seeding. In the Jaiba project in Brazil, farmers will be able to pump from their individual reservoirs. In the Westlands Water District, water must be ordered 24 hours in advance of deliveries, but can be shut without advance notice. Unfortunately few systematic studies have been done to establish a positive correlation between flexibility in irrigation delivery and crop yield. Indications are that there is a strong positive correlation has been observed by J. Merriam (1987, 1991a, 1991b) in Sri Lanka and India. Interestingly, the design objectives of most of the following rehabilitation projects were to increase reliability and operational flexibility.

The examples of the Manicoba and Jaiba-Mocambinho projects in Brazil present interesting examples of the versatility of modern technology. Perhaps even more striking are the examples of the Tepalcatepec-Cupatizio project in Mexico and Gadigaltar in India. In both cases, the construction of conventionally designed schemes had already started when modernization of the projects was decided. In the Mexico case, the designer had to race with the contractor to modify the construction drawings.

The presented designs concepts cover the full range from basic hydraulic structures to local automatic equipment and advanced automation techniques with remote control through central decision support systems and modern communications equipment. It is not the specific equipment but the design approach and the recognition of the operational requirements that characterizes modern irrigation.

Flow Dividers and Buffer Reservoirs

The projects of Peru, Sudan and Nigeria have been selected to illustrate the impact of simple but appropriate technology on water management. These projects use either the concept of proportional flow division or night storage, which makes them more manageable than extensively gated systems.

A part of the La Joya project in Peru is equipped with flow dividers which provide an equitable distribution of water between sub-sectors. The described concept of flow division has some similarity with the proportional allocation concept used in South Asia, but the flow divider placed on unsubmerged structures are more accurate than submerged fixed outlets.

The Gezira scheme in Sudan uses simple technologies by present-day standards. It was designed before the development of modern canal water-control technologies. The main branch and major canals are designed in accordance with the regime theory, the minor canals (equivalent to laterals or secondaries in other projects) are oversized for storing water at night. The introduction of interim storage resolved the problem of low efficiency during night irrigation and provided a good solution to the problem of matching water releases at the headworks and at critical points of the systems with demand in an upstream controlled system. A negative feature of the design is that the oversized minor canals act as very efficient silt trap.

Local Water Level Control

Local upstream and downstream water level control or a combination of these two methods are used in a number of countries. Automatic devices such as float-operated gates, flap gates, long-crested weirs, or composite regulators (a combination of weirs and undershot gates) are used for upstream control. For downstream control mainly hydraulic gates are used.[3] In the United States electrical gates, powered by solar panels and controlled by microprocessors, are increasingly popular. These gates can be remotely changed from upstream to downstream to flow control. It is also possible to change the set points.

The key features of the Kano project in Nigeria are buffer reservoirs for flexible water delivery and automatic flap gates (Begemann gates) that provide upstream water level control and ensures constant flow at offtakes.[4]

In some cases, reservoirs are incorporated in the system, for example to serve as the transition between upstream and downstream control (Gadigaltar, India; Doukkala, Morocco; Coachella District, United States) or to increase flexibility of water delivery without spills in upstream-controlled canals (Tepalcatepec, Mexico). In most of these

---

[3] The only manually operated downstream control in the Office du Niger project is not discussed here.

[4] Regrettably, the only known applications of this simple and efficient device are the Kano project in Nigeria and the Yaque del Norte in Dominican Republic. The gates require considerable head at the structure for proper operation but there are projects in a number of countries where this is not a constraint (projects in hilly areas of Indonesia, Philippines, Northern Thailand).

projects, flows at offtakes are controlled by modular distributors ensuring constant releases despite variations in the upstream water level.

Modular distributors have been widely used in North Africa since the 1950s. There have also been large scale applications in the Middle East (Iraq, Iran) and in a few projects in a number of other countries (West Africa, Madagascar, Malaysia). This technique is now being introduced in Mexico, Indonesia, and Thailand through pilot projects of 5,000 to 20,000 ha each.

Distribution

In most cases, water is delivered to the farm through a network of field channels for gravity irrigation. However, there are some exceptions. In the Kirkuk project in Iraq, the secondary canals serve networks of low-pressure pipes. The same concept has been adopted for the Gadigaltar pilot project in India. Farmers in the new Jaiba project in Brazil will pump from individual reservoirs or directly from secondary canals into pressurized on-farm systems. In the California districts, the distribution system consists of open pipes with overflow stands (Coachella), closed pipelines (Westland Water District) or semi-closed (Orange Cove).

Cost Comparison

Two projects allow the cost comparison of conventional and modern design: Tepalcatepec-Cupatizio in Mexico and Sidorejo in Indonesia. In both cases, the additional cost of modernization of water control is about 3 percent of the total cost of a conventionally designed project.

10. CASE STUDIES

The projects discussed are presented in alphabetic order of their country, to avoid the impression of any technical preference.

Brazil

About 90 percent of the 2.7 million ha now irrigated in Brazil has been developed by the private sector. Most of the irrigated lands developed by the public sector are located in the northeast, mostly in the Sao Francisco Valley. Public systems implemented in the 1970s were designed for gravity application and equipped with conventional, manually operated gates. Shortcomings in original design, mainly the absence of water control and measurement structures as well as low construction standards, aggravated by inadequate farming practices and unsatisfactory maintenance, have led to wide spread problems of unreliable water supply in some areas and waterlogging in others. Microtopography is often irregular, and soils are relatively shallow, a situation not favorable for land leveling and gravity irrigation. Agricultural production in those old public schemes is low and generally below the potential.

The two cases presented in the following illustrate how the modern canal operation concepts and modern irrigation technology can be applied to the modernization of old schemes. These projects were designed by Brazilian firms that developed innovative solutions within constraints imposed by existing structures.

Jaiba Project. The Jaiba project, financed by the World Bank, is the first phase of the largest public irrigation scheme under implementation in Brazil. The basic infrastructure of the scheme was constructed in the 1970s and designed to supply water to an area of 100,000 ha. Existing works include: (1) a pumping station with a capacity of 80 m^3/s on the Sao Francisco River; and (2) a lower main canal conveying water to a second pumping station and a short section of the upper main canal. The distribution system has never been fully completed. Only a gravity irrigation area of about 1,000 ha called Mocambinho was completed in 1987. Another area of about 1,700 ha, sector F, using sprinkler irrigation and served from the upper main canal, was put into operation in 1989. Microtopography of the project area is irregular and soils are generally light. Sprinkler and drip irrigation are the most suitable methods of irrigation.

Under the new project an area of about 17,800 ha will be equipped with sprinkler irrigation and the Mocambinho area will be converted from gravity to sprinkler irrigation. The operational objective is to provide water delivery service as flexible as possible.

Beneficiaries in the new area will be smallholders (6 ha each) and some medium and large farmers. Based on economic and other considerations, it was decided that water would be delivered by gravity to individual farms, each one equipped with an electric pump in a small sump. The secondary canals would be operated under downstream control to allow free demand delivery. Two alternative solutions were considered: (1) conventional top-level canals with float-operated gates, and (2) constant-volume operation (BIVAL type) with remote centralized control. Cost analysis of these two solutions gave preference to local automatic control.

The consultants have also studied various alternatives to convert gravity irrigation to sprinkler irrigation and free demand in the Mocambinho area. The conversion to downstream control is too expensive because of the presence of drop structures and steep slopes of existing canals. The consultants thus proposed upstream control with long-crested weirs and 150 individual reservoirs. Simulation studies determine the size of these reservoirs in the range of 100 to 500 m^3 each depending on the size of the farm and their distance from the secondary offtake.

Manicoba Project. The existing Manicoba project is on the right bank of the Sao Francisco River, 30 km downstream of the Sabradinho storage reservoir.

This scheme has a net service area of 4,400 ha of which about 2,400 ha are currently under irrigation. About 1,900 ha are occupied by small farmers who have an average farm size of 8 ha. The remaining part was allocated to large farmers. The main pumping station has a total capacity of 6.8 m^3/s. A 3 km long steel pipe serves the small farm area, and a 5 km pipe supplies a 28 km long canal that serves mostly the large farm area. Two booster pumps at the main canal serve small areas out of the direct command. All canals, including the final distribution system, are concrete lined. All control structures and the pumping stations are manually operated. There are only three regulators and three Parshall flumes in the main canal and none in the secondary and tertiary systems. Under these conditions, efficient operation of the system is difficult to achieve. Overall water-use efficiency is likely in the range of 25 to 30 percent.

Under the modernization project, the main canal and the secondary canals will be operated under upstream control through long-crested weirs. The operation of the three pumping stations will be automatic and controlled by water level in upstream reservoirs having a capacity of 100 to 900 m^3.

To provide for more flexibility in operating the systems, in particular for off-peak power utilization, and to allow for the transmission time in canals, 10 off-line compensation reservoirs will be built in the upstream sections of the secondary canals. Volumes of these reservoirs range from 1,000 to 12,000 m^3 with a total of 63,000 m^3. Flows from the control reservoirs into main canals will be controlled by modular distributors in combination with downstream constant-level gates. The compensation reservoirs will be build on the side of existing canals near existing drop structures to take advantage of available head. Flow releases will be controlled as above. Fifteen long-crested weirs will be constructed in main canal and about 150 in secondary canals for upstream water level control. Tertiary intakes on secondary canals will be equipped with modular distributors. The water control method requires raising of the berms of the canal by about 0.10 to 0.45 m. For this purpose a concrete wall will be added to the top of existing concrete lining. The total costs of these works were estimated at US$ 2,340 per hectare. Of this cost, the hydromechanical equipment accounts for only 14 percent.

The reason for not converting the scheme to sprinkler irrigation, as in the Mocambinho sector of the Jaiba project, are the high cost of pumping the water from the river. Additional pumping for sprinkler application would make the project uneconomical. The modernization of the system will largely improve the distribution of water to small farmers and alleviate waterlogging in some areas. However, upstream control of the distribution system will require the preparation of irrigation scheduling and strict discipline, while in Jaiba, farmers will receive water on demand.

Egypt

The key feature of old irrigation systems in the Nile Valley and the delta is that canals are below field level and farmers have to pump the water to their fields using traditional sakias and, recently, mobile diesel-driven pumps. This situation requires less precise water-level control and reduces overirrigation by farmers (since they pay for the cost of pumping). Access to water is generally continuous and reasonably reliable. The main and secondary canals are operated under continuous flow and upstream control with little or no variations in flow rates over long periods. Only the tertiary canals are operated on a seasonal adjusted rotation schedule. The system is operated with very few measurement points since no water charges are levied. Water management is based on observing spill at the downstream end of canals. Excess water is usually drained back into the Nile, but in some areas it has also caused waterlogging and salinization problems.

As long as pump sizes were small, there were few operational problems and farmers enjoyed a flexible water supply at times when their tertiary was "on". As farmers began to use larger pumps and as water became scarcer, problems such as water shortage and unequitable water availability appeared at the tail ends. The Egyptian government, supported by World Bank and USAID, is now examining new control options in anticipation of future water shortages from the Nile.

An Irrigation Improvement Project includes eleven individual systems distributed among 6 governorates and serving a total of 240,000 ha. To cope with the present deficiencies, the proposal has been to change the operation method of tertiaries from rotation schedule to continuous flow while maintaining on-demand deliveries, with improved flexibility and equity.

To achieve these objectives downstream control will be applied in the distribution canals and additional storage will be created to compensate for differences between nearly constant flow supplied by the principal and subprincipal canals under upstream control and the daily variations in demand from the users. In addition tertiary canals will be raised and individual pumps replaced by single point pumps for better distribution and lower pumping costs.

India

Substantial differences in agroclimatic and socioeconomic conditions and historical development of irrigation between regions of India have resulted in a wide variety of irrigation infrastructure and management concepts. A broad distinction can be made between systems that distribute water according to preestablished rules (water duties) and systems that attempt to meet crop water needs. These two broad categories are often referred to as supply-based and demand-based. However, in both cases, the irrigation scheduling is decided by the irrigation agency.

The following describes two pilot projects with different concepts. Majalgaon is part of a large-scale project in the state of Maharashtra and Gadigaltar is a small tank project in

Madhya Pradesh, where the farmers would have more responsibility in water allocation and distribution.

The Majalgaon Irrigation Project, located in the valley of the Godaveri River and serving 58,000 ha, is an expansion of the large Jayakwadi project with a total irrigable area of 350,000 ha. The expansion became possible after the Majalgaon storage dam, with a capacity of 450 million m^3, had been constructed to regulate the water of the Sandphana River. This water source is complemented through releases from Paithan dam into a link canal.

The Majalgaon main canal and parts of the distribution system had been partly built when the Irrigation Department of Maharashtra decided to use the project to test alternative delivery method to the shejpali system, commonly used in the state of Maharashtra. Under the shejpali system, individual farmers receive individual water allocations for sanctioned crops. However, the shejpali system requires a precision and flexibility of water control that is increasingly difficult to achieve with the existing control structures and the present fragmentation of land. The proportional delivery system, used in North India, is inflexible and not well suited to the conditions in Maharashtra, such as variable soil conditions, diversity of crops, and rolling topography. Under the new concept water will be sold in bulk to water user associations, each serving about 300 to 400 ha. The WUAs will get annual quotas and are responsible for the distribution of water and the maintenance of the water courses and field channels. They order the volume of water required for each irrigation turn. The internal water distribution will be essentially on rotation and proportional to the size of the holding, but other arrangements such as buying and selling of water are possible.

Several alternative water control methods were studied in 1990 to provide the required operational flexibility. However, since most of the system is already completed or under construction, the structural modifications had to be limited. The proposed solution makes use of the storage capacity of the main canal, which was originally sized to serve an area of 100,000 ha. The main canal will be operated under the constant volume concept. The cross regulators will be manually operated on the instruction of a central control center. The turnouts of branch and minor canals will be equipped with downstream control gates and modular distributors. Long-crested weirs and modular distributors will be installed in minor canals when the accuracy of existing structures is insufficient.

A preliminary estimate indicates that the cost increase caused by the addition and modification of control structures is on the order of 5 percent. Construction or modifications of these control structures is scheduled to start in 1993.

The Gadigaltar tank project in Madhya Pradesh has recently been constructed. The guiding criteria of this project is to provide farmers with control at his farm turnout on the frequency, rate, and duration of flow with some restrictions from the project organization in case of drought. Using the limited rate, arranged schedule, the small irrigation groups will be responsible for arranging with the farmers the irrigation schedule. It is anticipated that delivery will be made within one day of the water request. It is expected that the increased efficiencies through improved control and use of pipes will provide appreciably more water.

The project covers about 1,150 ha with more than 500 farms. The main infrastructure includes:

- A main storage tank with a total capacity of about 9 million m^3.
- A 5.5-km-long sloping canal from the main tank to a balancing reservoir. The canal is operated on upstream control with an initial capacity of 1.3 m^3/s.
- A balancing tank with a capacity of 80,000 m^3.
- A 1.5 km-long, top-level canal from the balancing reservoir to the lower irrigation area. This canal is operated on downstream control with automatic constant-level control gate.
- Semi-closed piped distribution system with outlets at every farm, taking off from the sloping and the level-top canals.

About 500 ha are served from the sloping canal through five pipe laterals and the remaining 660 ha from the level-top canal through five conveyance pipelines and one direct outlet. Float valves and an overflow stand are used to limit the static pressure in the pipes to about 5 m.

There are 67 groups of farmers, each with a group irrigator who arranges water deliveries with the central water coordinator's office. Most of the groups consist of 5 to 10 farmers with a total area of about 10 to 25 ha. Each group has its own water meter and a connecting gate to the supply system. The water will be supplied with a limited rate, arranged schedule with the ensured minimum flow rate of 30 liters per second.

Interestingly, construction of the primary tank and of the first 5 km of the main canal was nearly completed when the modernization of the design was decided. This constraint was solved through the construction of the overnight storage reservoir. For a detailed description see J. Merriam (1990).

Indonesia

Small run-of-the-river systems and small and fragmented land holdings are characteristic of Indonesian irrigation. Water is generally distributed under central control. Water demand for each tertiary block in an irrigation system is assessed by the administration every 10 to 15 days. This requires knowledge of actual irrigated areas, the type and area of each crop in every block, and precise estimates of conveyance losses. In practice, there is considerable reliance on estimated rather than precise information.

The physical infrastructure that has evolved in association with this management system requires a high degree of control. However, valid design standards recommend undershot gates at cross-regulators and adjustable Romijn weirs at canal intakes. This is the worst combination of control structures with regard to hydraulic stability, as explained in Part I of this paper. Recent studies by IIMI have shown that even if the operating procedures are truly followed, planned and actual discharges often differ considerably in practice.

Sidorejo Irrigation Project. In the mid 1980s, the Indonesia Directorate General of Water Resources Development selected the Sidorejo Irrigation Project, a subsystem of the proposed Kedung Ombo multipurpose dam and irrigated scheme to test modern canal control and to determine its replicability to other irrigation systems. The recently completed project is located in central Java about 70 km east of Semaranj and serves an area of 5200 ha. This project is of particular interest since it is the first large-scale experience in south and east Asia under predominant rice crop that has adopted an advanced control concept for an entire system.

Water is supplied to the Sidorejo pilot system through a diversion weir, approximately 9 km downstream of the storage dam. The main canal has a total length of 13 km and a maximum design capacity of 9 m^3/s. The main canal is operated under downstream control; the secondary and tertiary canals taking off from the main canal are operated under upstream control. The main canal is equipped with automatic float-operated gates maintaining a constant downstream level. The water level in the secondary canals is upstream controlled through automatic hydraulic gates. Discharges are controlled by modular distributors.

The control concept adopted for the Sidorejo pilot project is quite similar to the one used in Mediterranean countries. However, it differs in the control structures for the distribution system by not making use of static structures such as diagonal or duck-bill weirs. This deviation was justified at design stage by the risk of siltation but could create a maintenance problem in the future.

Hydraulic tests of the Sidorejo project have recently been completed. The main canal is operating satisfactorily, but several corrections were required in the installations of mechanical equipment in the smaller canals. The intent was to monitor the performance of the new design to determine the advantages and disadvantages over conventional design systems in Indonesia. The new design would allow change in the present water allocation procedure, which is based on historical rainfall records, to a real time operation based on actual rainfall measurements.

A detailed comparison of construction costs has been carried out by consultants. This study shows that the costs of the Sidorejo project with automatic control is similar to the cost of the project if it had been built using conventional Indonesia design standards. The hydromechanical equipment is more expensive, but these extra costs are compensated by substantial savings in civil works. In general, earthworks and lining are more expensive for canals under downstream control because of the need for horizontal banks. In Indonesia, however, a wide shallow section is adopted for traditional upstream-control canals to minimize variations of flow levels despite substantial variations in discharges. According to Indonesia design standards the ratio of bed width to water depth at maximum discharge should be 5 to 5.5. In the new design the ratio is only 1 resulting in a 16 percent cost saving. Using 1986 unit costs, the results of the costs comparison are presented in Table 10-1.

Table 10-1. Cost comparison

	Traditional System (Rps million)	Automatic System (Rps million)	Difference (percent)
Main canal			
Earthwork and lining	2,380	2,047	-16%
Flow control structures			
-civil works	165	133	-24%
-equipment	236	474	+49%
Other structures (roads, bridges, culverts)	2,347	2,231	-5%
Total	5,128	4,885	-5%
Secondary canal			
Earthwork and lining	134.5	134.5	0%
Flow control structures			
-civil works	11.1	11.1	0%
-equipment	15.6	22.9	31%
Other structures (roads, bridges culverts)	118.4	118.4	0%
Total	279.6	286.9	3%

Source: SOGREAH (unpublished).

Madagascar

Madagascar has no uniform national standards for the design of irrigated projects. Clear evidence of this is in the variety of approaches to water control in the conveyance and distribution system used in medium-sized schemes in the Lake Alaotra, which may reflect more the particular preference of the expatriate consultants than operational necessity.

PC 23. The design of PC 23 (9,200 ha) makes use of the advanced hydro-mechanical equipment for canal control: upstream and downstream constant-level gates and modular distributors. The scheme is composed of two run-of-river systems: a lower one with its main canal operated under upstream control and an upper one with a main canal operated under downstream control. Secondary canals are equipped with composite cross-regulators comprising of a short static weir section associated with a sliding gate. Turnouts for tertiary canals are equipped with modular distributors. Offtakes on tertiary canals are equipped with division boxes that have vertical dividing flaps to make possible variations in the sharing of flows between quaternary canals.

Sahamaloto. This scheme of 8,000 ha is supplied directly from a storage reservoir on the Sahamaloto River. The flow in the main canal is subdivided into three subsystems downstream of the dam through a structure equipped with three adjustable overshot gates (Romijn). All control structures in the system are of the Romijn type.

Anony. The run-of-river scheme (6,000 ha) combines the control structures used in PC 23 and in Sahamaloto. All cross-regulators are equipped with Romijn gates and the offtakes with modular distributors. Because of large variations of water level upstream of the diversion dam, the offtakes of the two main subsystems, one in each bank, have downstream constant-level gates followed by a battery of distributors.

The PC 23 scheme should be the simplest and easiest system to operate. Water levels are automatically controlled, and flows diverted at tertiary offtakes are stable between two settings of the module distributors. The Sahamaloto scheme is designed for operation on a proportional basis (with rotation at farm level). It is not clear what the objectives of the designer of the Anony scheme were when combining modular distributor offtakes with overshot gate cross-regulators; simpler composite regulators (long-crested weir with a simple gate) would have met the same objective. The variety of control concept and equipment in a small area confuses the operation staff and makes the transfer of management to water user associations difficult.

Mexico

Most of the irrigation canal systems serving the 77 large-scale irrigation districts in Mexico are operated under conventional upstream control and arranged water delivery schedules. The farmers can order water 72 hours in advance in some districts and one week ahead in others. The cross-regulators are, in general, equipped with sliding or radial gates; large regulators on main canals are motorized. All structures on lateral and sublateral canals are manually operated. Farm turnouts are equipped with single gates or, in some districts, with constant-head orifice outlets. Despite the limitations of conventional design, the systems in the arid northwest coastal region of Mexico are performing relatively well, thanks to good management, the motivation and training of operations staff, and the extensive, all-weather road system and radio communications network. In the more humid zones of the country with irregular topography and erratic rainfall, performance of the irrigation schemes is less satisfactory.

This section discusses two cases of modernization of canal systems with the objective of providing improved irrigation service to farmers:

1. The partial modernization of one of the Canal Alto of the Yaqui District in the northwest region.

2. The modernization of the right bank of the Tepalcatepec-Cupatizio project in central Mexico. The change of the concept was decided after the design based on traditional criteria had been completed and construction had started.

Yaqui Irrigation District, Canal Alto. The Canal Alto, with a capacity of 110 m^3/sec, serves about half of the 220,000 ha of Yaqui Irrigation District. Its total length of 125 km is divided in 18 sections by 17 cross-regulators, all equipped with manually operated, motorized gates. Two medium-sized, on-line reservoirs are part of this main canal. Farmer may order water twice a week for the next three to four days but control of water level and flow is difficult. Several studies were made over the last 15 years to improve water control. The project presently under implementation consists of converting from upstream control to the constant-volume concept. Results are not yet available.

An alternative modernization project considered was to maintain the upstream-control concept and water-scheduling process but to use a computerized decision support system to calculate the gate setting. This alternative was not implemented because it would not significantly improve the operation of the system.

Tepalcatepec-Cupatizio District: Chilatan Right Bank. The Chilatan irrigation system is part of the Tepalcatepec-Cupatizio Irrigation District in Michoacan state, about 300 km west of Mexico City. The 50 km long right main canal has a capacity of 25 m^3/sec and will serve about 20,000 ha. It has a gentle slope and is divided into seven reaches through six cross-regulators. A main lateral with a capacity of 11 m^3/sec branches at a distance of 12 km from the diversion dam and runs with a steep slope to serve about 50 percent of the project area.

The engineers had the challenging task of changing the control concept of the entire scheme with minimum modification to the control structures already completed and the construction drawings already submitted to the contractors. The solution was to convert the control method of the main canal from upstream control to controlled volume (BIVAL). This requires that the banks of only half the length of each reach are raised. It was achieved through the construction of concrete wall of a maximum height of 0.8 m. The branch canal will remain under upstream control but water level stability will be improved by replacing the four cross-regulators with composite regulators: at each regulator one or two vertical gates are to be removed and replaced by long-crested weirs. All constant-head orifice (CHO) turnouts will be replaced by modular distributors constructed of fiberglass and manufactured in Mexico.

To eliminate all possible operational losses, operation of the branch canal will be further improved through an off-canal compensation (buffer) reservoir and in-line extra capacity through raising the canal lining in four selected sections. Operation of the main canal under BIVAL and selected cross regulators of the branch canal will be done from a central location via cable communications.

Detailed cost analysis has shown that some cost can be saved through the modification of the control structures on the main lateral. The cost of the long-crested weirs of the branch canal is offset by the savings of one or two gates on each composite regulator. Fiberglass distributors are less expensive than conventional steel gates. The additional costs are limited to:

- Raising half the length of the main canal by 0.0 to 0.8 m.
- Raising some sections of the main lateral.
- Remote control center, communication system and motorization of gates.

Overall, it has been estimated that the cost of design modifications would not exceed 5 percent of the total cost of the entire system. The expected benefits from the modernization of this project are:

- Elimination of all operational losses from the conveyance and main distribution system.
- Automatic operation of the main conveyance system, which will reduce the advance time required to prepare and adjust deliveries schedules.
- Potential application of volumetric water charges and volumetric allocations.

Middle East

There is virtually no information available in the irrigation literature on the performance of large-scale modern irrigation schemes built in the Middle East in the 1960s and 1970s. This section discusses some aspects of two large schemes featuring modern design concepts. The first scheme, near Kirkuk in the north of Iraq, covers 230,000 ha and is still under construction, the other scheme is located in the Guilan plain in northern Iran near the Caspian Sea, it covers 270,000 ha and was completed in the late 1960s. An in-depth performance study of the Guilan project which has now been in operation for more than 20 years, would certainly provide very useful lessons on the modernization of irrigation schemes.

Kirkuk Project, Iraq. Construction of the Kirkuk project about 300 km north of Baghdad proceeded rapidly until the mid 1980s when it stopped because of political and military events. The project is supposed to divert water from a tributary of the Tigris River, regulated by the Dokan dam with a storage capacity of 5 billion m^3. The climate of the region is continental, and the annual average rainfall is about 350 mm. The soils of fine texture are moderately saline, but contain a high proportion of gypsum, which pose serious problems for canal lining.

The project comprises essentially of:

- A 37 km long feeder canal from the diversion dam to the head of the main canal with a design discharge of 280 m^3/s.
- A 95 km long main canal with a design capacity of 230 m^3/s.
- Secondary canals serving networks of low-pressure pipes supplying hydrants at the farm turnouts. The total net irrigated area will include 200,000 ha of gravity irrigation and 30,000 ha of sprinkler irrigation.

To ensure minimum response time and high flexibility in water distribution, the feeder canal is regulated by the associated levels remote control method, maintaining constant head loss between two successive regulators. Each regulator is automatically controlled and monitored by a central computer, communication is provided through radio links. Stability of the regulators was checked on a mathematical model. The gates of the main canal are also automatically controlled, using P.I.D. controllers.

The flows in the secondary canals are controlled by hydraulic downstream control gates and modular distributors. Slide gates control the pipe offtakes on the canals. The hydrant discharges at the farm turnouts, fixed at 70 l/s, are controlled by flow limiters.

Guilan Project, Iran. The Guilan project located on the southeast coast of the Caspian Sea benefits from a particular climate unique in western Asia. The region is open to maritime influences and isolated from the Iranian plateau by the Elbrouz Mountains, which rise to elevations ranging between 2,500 and 3,000 m. Temperatures are similar to the Mediterranean climate, but average annual rainfall exceeds 1,000 mm. The project is supplied by Sefi Roud River water regulated through a large storage dam with an active capacity of 800 million m^3. There are considerable variations in the annual flow of Sefi Roud, with a median flow of 3.6 billion m^3. Rice is the main irrigated crop; average rainfall during the growing season between April and September is about 250 mm. The area is densely populated (more than 330 habitants per km^2), and the average farm size is 0.8 ha.

The project implemented between 1962 and 1969 comprises two irrigation areas, each supplied from a diversion dam: (1) the upper zone covering 70,000 ha with a 17 km long tunnel and the 52 km Foumen long canal, with a capacity of 35 m^3/s, and (2) the lower zone covering 81,000 ha on the right bank and 115,000 ha on the left bank served by two main canals of 67 and 114 m^3/s respectively.

The two main systems are operated under upstream control through use of long-crested weirs and hydraulic gates. The offtakes are equipped with modular distributors. About two-third of irrigated area (243,000 ha net) are still served by traditional unlined canals. The rest of the area is served through lined branch canals and equipped with long-crested weirs and modular distributors and raised prefabricated canals (canalettis). The civil works are still in excellent condition, and the hydromechanical equipment is progressively replaced after 25 years of service.

Water releases are decided by the Guilan Water Authority, and water users are informed through their water masters, who operate the secondary systems and report on the system status back to the Water Authority. On average, 1.7 billion m^3 are diverted annually to irrigate about 142,000 ha of paddy fields. It seems that the project performance in terms of efficiency is reasonably good and close to the expectations at the planning stage. Operation of the project is simplified through the use of hydraulic regulating structures and appears to be quite appropriate for the prevailing rice irrigation with moderate rainfall during the growing season.

Malaysia

The initial stage of irrigation development in Malaysia was intended to provide controlled drainage to existing rice lands. With the advent of double cropping after independence in 1957, the main objective was the development of reliable water supply for the second crop, which involved the construction of storage and diversion dams and the upgrading of existing irrigation systems. This section describes two large irrigation schemes with aspects of modern water control.

Muda scheme. The scheme of 98,000 ha is located in the northwest of peninsular Malaysia. This scheme accounts for 40 percent of the national rice production and is critical to the rice policy. The project is not only a case of successful upstream control, but it is also one of the best documented projects. The infrastructure comprises of two storage reservoirs connected by a tunnel, a diversion dam 35 km downstream, and two main canals, north and south, running along the perimeter of the command. Cross-regulators on these canals are equipped with over-shot motorized gates. Offtakes serve the lateral canals that run westward to the sea and the sublaterals, typically at 2 km intervals. Cross regulators on lateral and sublaterals as well as offtakes, are simple, hand-operated, undershot structures. A remote monitoring system in project headquarters was built to provide engineers with real time information on reservoirs and canal water levels and on rainfall in the catchment areas between the storage and diversion dams to predict the unregulated flow.

The combination of remote monitoring system and overshot gates on the main canals has contributed to the efficient operation of the main system. The adoption of conventional water control for the distribution system, however, resulted in fluctuating water levels and unreliable water deliveries to rice growers.

The mechanization of rice farming, particularly the conversion from transplanting to direct seeding, had marked effects on demand for irrigation water. It is now much more important for farmers to control the amount and timing of water deliveries. In practice, it also meant that planning and harvesting dates are less uniform than they used to be. To achieve the control over amounts and timing of water delivery, farmers in Muda have installed their own low-lift pumps to lift water from public canals and drains. Although the costs are higher than for public irrigation water, pumping is preferable to insufficient and untimely water supply and subsequently reduced yield.

Kemubu-Kemasin. Improving the operation of the distribution system through easily operated structures is the approach adopted for the Kemasin extension of the Kemubu project.

The Kemubu project is located opposite the Muda project on the northeast of peninsular Malaysia near the Thai border. The 17,000 ha project was originally supplied by pumping Kelantan River water through 5 diesel-driven pumps with a total design capacity of 30 m^3/s. The south main canal is operated under downstream control. A few of the lateral canal offtakes are equipped with modular distributors and the remaining by adjustable Romijn type weirs. The farm intakes are equipped with adjustable undershot gates combined with a flow-measuring device. After completion in mid-1975, some difficulties occurred when it was found that the effective capacity of the pumping station was 30 percent less than expected and that the proportion of the light soils was higher. This problem has been solved with the completion in 1989 of a new pumping station of 6 electrically driven pumps with a total capacity of 42 m^3/s.

Another problem in Kemabu was to control the flow in the minor distribution system through manually operated gates and the increasing importance of diversified cropping, that required fine tuning of water supply. These difficulties may have been the reasons for selecting a different control structures for the Kemasin area extension of the project. The distribution system for that extension is equipped with gated flow-dividing structures making it possible to divide the incoming flow either proportionally to areas served (when all gates are open) or according to farmer's requests (flexible structured design).

Morocco

The development of modern control methods using automatic hydraulic devices started in North Africa in the late 1940s following an intensive hydraulics research program in France. The use of these devices became widespread in Algeria, Morocco, and Tunisia in the 1950s with Morocco, having the largest irrigated area of the three countries, taking the technological lead. The large-scale schemes cover 380,000 ha or about 45 percent of the total area (900,000 ha). Most of these schemes have the following characteristics:

- Farm sizes are generally small, between 2 and 5 ha.
- Water is delivered on arranged demand schedule to individual farmers.

- Delivery is by volume and water charges are calculated volumetrically.
- The main canals are operated under upstream or downstream control or a combination of the two methods, and the distribution system is operated under upstream control. Dynamic regulation, a central control method, is used for the 110 km main canal of the Haouz project near Marrakech, which was installed in 1990.
- The main canals are equipped with automatic hydraulic gates and the distribution system with long-crested weirs. The canal offtakes are equipped with modular distributors. The farm offtakes are equipped with distributors, flow dividers, or on-off gates.

The control concept of two large schemes is briefly discussed below.

Doukkala project. The upper half of the main canal is under upstream control and the lower half is under downstream control with top-level canals. A regulating reservoir at the midpoint of the main canal acts as a buffer for the variations in flow rate. Releases from the storage dam are simply based upon the next day's cumulative demand and adjusted depending on whether the buffer reservoir is filling or emptying. The small storage capacity of this reservoir has been able to absorb variations between supply and demand. But the development of sprinkler irrigation on a demand basis in the area served by the lower half of the main canal will require further modification of the control method.

Tadla project. Two irrigation canals start from the afterbay of a hydropower plant that releases water daily during peak hours. Since the capacity of the afterbay is far too small to store the volume of water released during peak hours, special gates have been installed to store the excess flow in these two canals while maintaining the downstream control operation.

The experience of Morocco and Tunisia, shows that it is possible to deliver water to small farmers on the basis of prearranged demand and to apply volumetric water charges. This is feasible thanks to the use of reliable, easily operated structures and, to some extent, to the consolidation of small, fragmented holdings.

Nigeria

Kano River Project. The Kano River project (23,000 ha for the first phase) in northern Nigeria is one of the largest irrigation projects, not only in that country, but in West Africa after the irrigation schemes of the Office du Niger in Mali. The irrigation water is conveyed from the Tiga storage dam to the irrigated area through an 18 km canal that splits into two main branches. The entire system is operated under upstream control. The design of this project is particularly interesting. Using only simple hydraulic principles and control structures, good control of water levels and flows throughout the system has been achieved and inefficient night irrigation has been avoided. Water release schedules are updated weekly in response to farmers requests. The control structures of the project include:

- Seven night-storage reservoirs of 100,000 to 200,000 m^3 each at different locations. The reservoirs are filled up during the night and water is released during the day. The flow from the reservoir is diverted during daytime to areas which are not under the command of the storage reservoirs.

- Automatic flap gates of very simple but robust design, that maintain a constant upstream level, are used throughout the scheme in the main canals and the distribution system. The installation of these gates is possible because of the large drop across the regulating structures.

- Turnouts consisting of a slide gate (on-off) and a battery of semi-modular gates that operate under constant upstream water levels controlled by the flap gates.

P. Pradhan (1993) observed that the installation of automatic water control devices (Begemann gates) contributed significantly to the low level conflicts among water users. Only 15,000 ha have been developed and are irrigated at present but the construction on the remaining area is continuing. Although the agricultural production suffers from irregular supply of other inputs, the farmers achieve an average yield of 4 tons/ha for rice and 2.5 tons/ha for wheat, as a result of fairly reliable water supply.

Peru

La Joya Irrigation District. The Irrigation District in the coastal zone of southern Peru offers an excellent example of non-adjustable design and rotational water supply. La Joya district consists of two subdistricts: La Joya Antigua (3,850 ha) completed in the early 1940s and La Joya Nueva (4,625 ha) completed in the 1960s after the main canal had been extended. The project is supplied from the Rio Chili, which is partly regulated by three storage reservoirs in the Andes Mountains. Despite that regulation, there are daily variations of up to 20 percent in the flow arriving at the La Joya District because of the variable abstractions of upstream users for power generation, municipal water supply of the city of Arequipa, and the preference of farmers in the Arequipa area for day irrigation. The offtakes from the main canal to important laterals consist of flow dividers that divert a fixed ratio of the total flow to each lateral independently of the total flow. Some of the smaller offtakes are equipped with adjustable gates. Fixed flow dividers are also found in the lateral system. Some of them are equipped with gates to provide some flexibility in the water allocation. Water at sector level of about 300 ha is distributed on a 3.5 or 4.5 day rotation in normal years, and on a 6.5 day rotation during dry years. Duration of water delivery is proportional to the area of each farm. Night and day irrigation are alternated between farmers at each rotation.

Two factors contribute to the smooth operation of water distribution in the La Joya Irrigation District:

- The flow dividers function well because of their location on unsubmerged structures and stable canal sections, thus providing equitable water distribution to the two subdistricts and to sectors independent of the seasonal and daily variations of the flow diverted from the Rio Chili.

- Two strong water user associations, one for each subdistrict, are responsible for the management and maintenance of the project. The technical and administrative staff hired by each association is in charge of operation and maintenance activities and the bimonthly collection of water charges.

Some modifications to the infrastructure of La Joya District are now under consideration to reduce the perception of unequal water distribution of the farmers of La Joya Nueva and the potential risk of conflicts between the two associations. The planned modification would replace the adjustable offtakes by one flow divider and a link canal to supply these laterals. Other farmers are concerned with the inflexible and unpredictable water supply which makes efficient on-farm irrigation difficult. Three sectors of about 500 ha each are now being converted to sprinkler and drip irrigation through construction of lined reservoirs and piped distribution systems equipped with individual meters.

The water-management concept and the use of flow dividers of the La Joya irrigation district has some similarity with the warabandi system used in Pakistan and the northwest region of India. The operational objectives are similar: equity of water distribution and simplification of operation through nonadjustable structures. There are, however, some important differences. Because of the heavy sediment load of the rivers in India, the distributary and minor canals are designed on the regime theory and are supposed to be either closed or run at full capacity. There are few cross-regulators to control water levels, and the offtakes consist mainly of open outlets. Despite the valid design criteria, siltation, bank erosion and weed growth affect the water levels in these canals, resulting in unequitable water distribution. The lesson is that excellent maintenance is a precondition for non adjustable, structured design.

Sudan

The key design feature of the 800,000 ha Gezira project, the largest scheme in Africa, is this: the tertiary canals that feed the fields are oversized and serve as buffer reservoirs at night. The particular design of the tertiary canals provides a very flexible, "on-demand" operation of field turnouts during the daytime. The large capacities allow day-only irrigation without the associated loss of night flows typical on many projects.

In recent years, the farmers have taken over management of the tertiary canals and have switched to 24 hour irrigation. Now some tailender problems appear because some of the large laterals no longer fill up during the night and because of the neglected silt and weed control in the laterals.

Water distribution from the Gezira system to the fields is efficient, timely, and reliable as long as the system is adequately maintained. The design was able to adjust to a major change of the original operational plan thanks to the built-in flexibility of the design of laterals. The main drawback of this unique feature of the Gezira scheme is that the laterals work as very efficient silt traps[5].

---

[5] A detailed presentation of the Gezira project can be found in the World Bank Technical Paper No. 120.

Sri Lanka

Modernization of two distributary canals was carried out in Sri Lanka under the Major Irrigation Rehabilitation project financed by the World Bank. Although of modest scale (150 ha each), monitoring these two canals provides some interesting information.

The structural modification in the design of these two distributaries, compared with a conventional gated system, included:

- Automatic constant downstream-level gates associated with modular distributors at the head of the distributaries.
- Baffle distributors at farm offtakes.
- Flow dividers within the farm units.

Monitoring of these distributaries and of the areas served did not show significant differences in water use and crop yields compared with their respective control areas, but some doubts exist regarding the reliability of these data. It was found, however, that the operational costs of the new design were about 40 percent lower then the operational cost of the conventional design. The general feeling in the Sri Lanka Irrigation Department is that the modifications provide a superior operation facility. The farmers also express their appreciation for these controls as they can easily check the quantity of water delivered to their field.

J. Merriam reports in various publications (1985, 1987, 1991b) on a pilot project in Area H of the Mahaweli Scheme. The pilot project covers about 150 ha on the lower end of a unit served by one distributary. The farmers are provided water on demand, flow is limited to 20 l/s. The automated supply system consists of a reservoir, a downstream level control gate, top level canals, and low pressure concrete pipes supplying individual turnout valves. The objective of the project was to compare the demand schedule to a conventional agency controlled rotation schedule. The principal points of comparison were: costs, crop production, water use, and social effects. The project operated successfully for six seasons (three years) and interesting results were achieved. Unfortunately, after the end of that period the project reverted to rotational supply schedule like the rest of the area. No water user association had been developed to prevent this, though the farmer results had been very satisfactory and further experimentation would have been valuable.

United States

Coachella Valley Water District, California. This project, located in the desert of southern California, supplies 33,000 ha of irrigated land. Primary crops are citrus, grapes, dates, alfalfa, and vegetables. Rainfall is negligible. Farmers in some areas of the district have groundwater wells; most rely completely upon canal water deliveries.

A canal conveys the water over a distance of more than 230 km from the Colorado River. The conveyance canal feeds into a 60 km long, lined main canal. Manually operated offtakes supply a piped distribution system with a considerable grade. The distribution system is designed as an open pipeline system (more than 805 km of concrete pipelines, varying in diameter from 30 to 244 cm) with overflow stands (approximately 7 m high) to provide pressure on turnouts. The water is delivered on an arranged basis, with a 24 hour advance notice. Most deliveries are started, stopped or adjusted in the early morning when the district employee (Zanjero) arrives at the farmer turnout in response to a farmer request. All farm deliveries are metered volumetrically with propeller meters.

Much of the main canal is lined to prevent seepage. Water levels and gate positions are remotely monitored and controlled but manually operated using a modified form of upstream control from a central office. A large buffer reservoir at the end of the canal absorbs daily differences between water orders and deliveries. The buffer reservoir also supplies the pipelines at the lower end of the district. The ends of all pipelines are remotely monitored for spill so that adjustments of flow into the pipes can be made if spill occurs.

Because of the nature of open pipelines, the deliveries are rigid during the 24-hour delivery periods. Farmers have built numerous reservoirs on their property to increase the flexibility of irrigation, mainly for labor convenience. In many cases, the water can flow by gravity from the reservoir to the field. In other cases, the water must be pumped. On-farm irrigation systems are a mix of drip systems, sprinklers, and surface irrigation. Typical field sizes are 10 to 90 ha. Crop yields are excellent and farms are intensively managed. The professional staff reports to the manager, who is hired by and reports to the board of directors. The board of directors is elected by the landowners of the district.

Westlands Water District, California. Westlands Water District was formed in 1952 upon petition of farmers and landowners located within the district's proposed boundaries. The district obtains water from the U.S. Bureau of Reclamation (USBR), which operates a series of reservoirs, river releases, and canals to bring water from more than 500 km away. The San Luis Canal, operated by the USBR, runs through the district, with approximately one-third of the service area uphill from the canal.

Westlands covers 240,000 irrigated ha. Annual effective rainfall is approximately 120 mm. Soils are deep, but about one-third of the area (found in the lower elevations) is affected by a high water table. More than 50 crops are grown in the district, but major acreage is planted in cotton, tomatoes, almonds, vegetables, and small grains. Farm ownership is limited to 380 ha per individual; most fields are 16 to 64 ha in size. The majority of the area is irrigated with furrows, but there is increased use of hand-moved sprinklers and drip on row crops. Almost all of the orchards and vines are drip irrigated.

Water contracts provide for 0.70 m of irrigation water per year. This is not sufficient for complete irrigation of all land. Therefore, farmers carefully select and rotate crops to best utilize the limited water supply. The acreage above the San Luis Canal (part of the California Aqueduct) receives no district water during drought years. A persistent six-year drought has encouraged many farmers in the lower areas, where groundwater is available, to drill deep wells for supplemental irrigation water (even though it is of poor quality).

Every year, the district is advised by the USBR regarding the estimated volume of water that will be available, based upon reservoir storage and the snowpack in the mountains. District farmers then decide on the crops and acreage for planting. A farmer has complete latitude regarding when the water will be used; however, the volume of water available per year is fixed each year.

The district's distribution system is a closed pipeline facility, consisting of 1667 km of reinforced concrete pressure pipe, varying in diameter from 25 to 240 cm. Thirty-eight lateral pipelines are gravity fed, while water is pumped uphill into 27 laterals. Each field has a turnout on the uphill side. No tailwater (surface runoff) is allowed from any field. Each turnout includes a propeller meter providing information on both the flow rate and volume delivered.

Farmers receive water on an arranged schedule, with a maximum turnout flow-rate of about 180 to 300 l/s per 64 ha. Water must be ordered 24 hours in advance of deliveries. Farmers can open and close their own turnouts. Water can be shut off at any time without giving the district advance notice. Water costs are about US$40 per 1,000 m^3.

The pressures can vary significantly with time at the turnouts, due to changing flow rates in the laterals. The original design specified an automatic flow control valve at each turnout; they proved to be defective so they were later removed.

The district has an ambitious water conservation program that promotes good on-farm water management. On-farm irrigation efficiencies average about 70 percent, and the district irrigation efficiency is about 85 percent. The district efficiency is higher than the on-farm efficiency because much of the deep percolation on the upslope farms shows up as a high water table in fields at lower elevations and is consumed by the plants in those lower fields.

Imperial Irrigation District (IID). IID is located in the desert at the southern end of California. It services approximately 184,000 ha of irrigated land. Major crops include alfalfa (about 50 percent of the acreage), sudan grass, small grains, sugar beets, and vegetables. Crops are grown year-round. Field sizes average about 16 ha. Average water deliveries are 1.8 m of water. The district irrigation efficiency is 75 percent, including on-farm losses and conveyance losses. The whole district is supplied by canals, most of which are unlined. A system of surface drains is installed throughout the district, with an access point for each field.

Seepage losses and on-farm deep percolation losses are low because of the very high clay content of the soils throughout most of the district. Spill from the laterals and mains account for about a 5 percent loss, tailwater losses are about 10 percent, and canal evaporation and seepage account for about a 6 percent loss. Flow rates into the district and all spill points are rigorously monitored because of water conservation restrictions imposed by the state and federal governments. A program to recover the tailwater of furrow irrigation has been started recently.

All of the IID losses (except evaporation) eventually end up in the Salton Sea which is a salt sink. A recent agreement with the neighboring Metropolitan Water District has provided IID with US$ 150 million for a modernization program in return of the transfer of the saved water. Consequently, the district has embarked on an ambitious water conservation program which also improves the quality of the irrigation service. Major aspects of the program include:

- Canal lining to reduce seepage. The new lining, however, provides for more head (higher canal water levels) at the turnouts, thereby reducing the effect of fluctuating canal levels on turnout flows.

- Improved remote monitoring. A microwave system has been set up in conjunction with radio telemetry to monitor key water levels and flows throughout the district.

- Automation of key canals. Most of the automation is for upstream or flow control, but the communications system and electronics are designed for eventual local downstream control on sloping canals. New electrically operated overshot gates are being installed at some locations.

- Increased buffer reservoir storage. Several new buffer reservoirs are installed to provide flexibility with upstream control.

- Automation of lateral canals. The lateral canals are being provided with automatic upstream control, to stabilize flow through the turnouts. Solar powered, microprocessor-controlled overshot gates are increasingly used because of their operational flexibility.

- Lateral interceptors to prevent spill from the ends of the lateral canals. The spill is captured and used in other areas of the district, rather than being discharged into the drains.

- Improved flexibility of deliveries. Deliveries have always been made on an arranged basis, but now users can order water for less than 24 hours.

The district has embarked on a significant training program for its employees, with the intention of making them more "service-oriented". The result has been less conflict with farmers.

References

Abernethy, C.L. undated. *Performance Criteria for Irrigation Systems*, IIMI, Colombo, Sri Lanka.

Abernethy, C.L. (1991). *Indicators of the Performance of Irrigation Water Distribution Systems*, IIMI, Colombo, Sri Lanka.

ASTAG (1990). *Indonesia, Irrigation Sub-Sector Project*, Asia Technical Department, World Bank, Washington, D.C.

Berkoff, D.J.W. (1990). *Irrigation Management on the Indo-Gangetic Plain*, Technical Paper 129, World Bank, Washington, D.C.

Bhutta, M.N. (1989). *Effect of Varying Discharges on the Equity of Water Distribution in the Irrigation System*, Draft Ph.D. thesis, University of Engineering and Technology, Lahore, Pakistan.

Bos, M., H. Murray-Rust, D. Merrey, H. Jonhson and W.B. Snellen (1993). "Methodology for Assessing Performance of Irrigation and Drainage Management" presented at the ICID Congress, the Hague, the Netherlands.

Bottrall, A.F. (1979). *Comparative Study of the Management and Organization of Irrigation Projects*, (draft), World Bank Research Project 671/34, Overseas Development Institute, London.

Burns, R. (1993). "Irrigated rice culture in monsoon Asia: The search for an effective water control technology", *World Development*, Vol. 21, 5: 771-789.

Burt, C.M. and G. Gartrell (1991). "Canal Models and You." Proceedings of the 1991 National Conference of the Irrigation and Drainage Division of the American Society of Civil Engineers, held in Honolulu, Hawaii, pp. 238-243.

Burt, C.M. and H. Plusquellec (1990). "Water Delivery Control", in Hoffman, G., T. Howell, and K. Solomon (eds.), *Management of Farm Irrigation Systems*, American Society of Agricultural Engineers, St. Joseph, Michigan.

Burt, C.M., P. Canessa and J. Parrish (1991). *Irrigation Efficiencies in Parts of the Selenium Drainage Area of the West Side of the San Joaquin Valley*, Central Valley Region, California Regional Water Quality Control Board, Sacramento, California.

Burt, C.M., J. Gartung, J.M. Lord, J. Parrish and J.C. Bryner (1981). *Distribution System Improvement to Facilitate Water Delivery*, A report prepared for the Office of Water Conservation, California Department of Water Resources, by J.M. Lord, Inc.

Bybordi, M. (1989). "Problems in Planning of Irrigation Projects in Iran", in Rydzewski, J.R. and C.F. Ward (eds.), *Irrigation: Theory and Practice*. Pentech Press, London.

Clement, R. (1965). *Le Calcul des Debits dans les Reseaux d'Irrigation Fonctionnant "à la Demande*, Societé du Canal de Provence, Le Tholonet, France.

Clemmens, A.J. (1987). "Delivery System Schedules and Required Capacities", in Zimbelman, D.D. (ed.), *Planning, Operation, Rehabilitation and Automation of Irrigation Water Delivery Systems*, American Society of Civil Engineers, New York.

Feng, Shen and H. Wolter (1993). "The Four Waters Concept in China", *GRID*, Issue 3.

Freeman, D. and M. Lowdermill (1985). "Middle-level farmer organizations as links between farms and central irrigation systems", in Cernea, M. (ed.), *Putting People First*, Oxford University Press.

Goussard, J. (1992). *Automation of Canal Irrigation Systems*, Working group on Construction, Rehabilitation, and Modernization of Irrigation Systems, International Commission on Irrigation and Drainage.

Horst, L. (1987). *Choice of Irrigation Structures (the Paradox of Operational Flexibility)*, Wageningen Agricultural University, The Netherlands.

ICID (1989). *Planning the Management, Operation and Maintenance of Irrigation and Drainage Systems*, Technical Paper No. 99, World Bank, Washington, D.C.

IIMI (1989). *Efficient Irrigation Management and System Turnover, Volume 2: Efficient Irrigation Management*, IIMI, Colombo, Sri Lanka.

Jurriens, R., A.F. Bottrall et al. (1984). *Evaluation of Irrigation Design: A Debate*, Network Paper 9b, Overseas Development Institute, London.

Lele, S. and R. Padil (1993). "Working for farmers' participation in major irrigation projects", proceedings of the National Workshop on Farmer Participation in the Management of Irrigation Systems, Hyderabad, India.

Levine, G. and E.W. Coward, Jr. (1986). *Irrigation Water Distribution Implications for Design and Operation, Vol. I and II*, AGREP Division Working Paper 125 and 125a, World Bank, Washington, D.C.

Merkley, G.P. and W.R. Walker (1991). "Centralized Scheduling Logic for Canal Operation", *Journal of Irrigation and Drainage Engineering*, ASCE 117(3): 377-93.

Merriam, J. (1985). *Demand Irrigation Schedule, Concrete Pipeline Pilot Project, Mahaweli Development Board, Area H, Block 404, D-1*, Mahaweli Development Board, Sri Lanka.

Merriam, J., N.G.R. DeSilva, and D.J. Bandaragoda (1987). "Six Seasons of Demand Schedule Irrigation for Improved Water Management in Sri Lanka", International Commission on Irrigation and Drainage, Thirteenth Congress, Rabat, Morocco.

Merriam, J. (1990). *Galdigaltar Tank Irrigation Pilot Project*, Irrigation Department, Khargone, Madhya Pradesh, India.

Merriam, J. (1991a). "Flexible Irrigation Supply Pilot Projects. Part I Principles", ASCE I&DD Conference, Hawaii.

Merriam, J. (1991b). "Flexible Irrigation Supply Pilot Projects. Part II: Sri Lanka, Pakistan, India, Egypt", ASCE I&DD Conference, Hawaii.

Murray-Rust, D.H. (1990). *Selecting the Appropriate Management Strategy*, IIMI, Jakarta, Indonesia.

Murray-Rust, D.H. and W.B. Snellen (1991). *Performance Assessment Diagnosis* (draft), IIMI, ILRI, and IHE.

Nickum, J.E. (1982). *Irrigation Management in China: A Review of the Literature*, World Bank, Washington, D.C.

Operations Evaluation Department (1989). *World Bank Experience with Irrigation Development: Socioeconomic, Institutional, and Technical Impact and Lessons, Vol. I, II and III*, World Bank, Washington, D.C.

Operations Evaluation Department (1990). *World Bank Experience with Irrigation Development: Socio-economic, Institutional, and Technical Impact and Lessons, Vol. IV, V and VI*, World Bank, Washington, D.C.

Operations Evaluation Department (1991). *Malaysia, Northwest Selangor, Krian-sungei Manik, Muda II, Rompin-Endau Projects*, OED Project Performance Audit Report, World Bank, Washington, D.C.

Operations Evaluation Department (forthcoming). *A Review of World Bank Irrigation Experience*, OED Special Study, World Bank, Washington, D.C.

Patil, R.K. and K.R. Datye (1989). Discussion of "equity considerations in the modernization of irrigation systems" in the ODI Network Papers.

Perry, C. (1992). *Irrigation in Conditions of Water Shortage, the Case of India*, mimeo.

Plusquellec, H. (1988). *Improving the Operation of Canal Irrigation Systems*, A seven-part slide series. Washington, D.C.: The Economic Development Institute of the World Bank.

Plusquellec, H. 1989. "Two Irrigation Systems in Colombia; Their Performance and Transfer of Management to Users' Associations." WPS 264. Agricultural Production and Services Working Paper. World Bank: Washington, D.C.

Plusquellec, H. (1990). *The Gezira Irrigation Scheme in Sudan: Objectives, Design, and Performance*, Technical Paper 120, World Bank, Washington, D.C.

Plusquellec, H. (1991). The Debate over the Size and Organization of Water User Associations in Irrigation Projects." Internal note of the World Bank.

Plusquellec, H. and T. Wickham (1985). *Irrigation Design and Management: Experience in Thailand and Its General Applicability*, Technical Paper 40, World Bank, Washington, D.C.

Plusquellec, H.L., K. McPhail and C. Polti (1990). "Review of Irrigation System Performance with Respect to Initial Objectives", *Irrigation and Drainage Systems*, Vol. 4:313-327.

Rey, J. (1990). *Field Application of the Kirindi Oya Right Bank Main Canal Simulation Model*, CEMAGREF, Montpellier, France.

Seckler, D. (1993). *Privatization of Irrigation Systems*, Discussion Paper No. 12, Center for Economic Policy Studies, Winrock International.

Shanan, L. (1986). *Design and Operation Guidelines for Structured Irrigation Networks*, South Asia Projects Department, World Bank, Washington, D.C.

Small, L.E. and M. Svendsen (1990). *A Framework for Assessing Irrigation Performance*, International Food Policy Research Institute, IIMI, and Rutgers University.

Tiffen, M. (1988). *Taking Social Economic and Institutional Aspects into Account during Project Planning*, Discussion Paper, ODI, London.

Uphoff, N. (1986). *Improving Irrigation Management with Farmers Participation: Getting the Process Right*, Westview Press, Boulder, Colorado.

U.S. General Accounting Office (1983). *Irrigation Assistance to Developing Countries Should Require Stronger Commitments to Operation and Maintenance*, GAO, Washington, D.C.

Valera, A. and N.B.J.M. Desa (1991). *Study on the Design-Management Interactions and Performance of Surface Irrigation Systems: The Kerian Irrigation Scheme, Malaysia*, (draft). Internal World Bank report.

World Bank (1993). *Water Resources Policy*, World Bank, Washington, D.C.

Methods of Operation and Structures for Main and Secondary Irrigation Distribution Systems

Operational concept	Main canal structure	Function of structure	Advantages	Disadvantages	Applications or design notes	Practice to improve operation	Project and country
1. FIXED UPSTREAM CONTROL	Proportional dividers	Divide incoming flow into predetermined (and generally fixed) proportions at each bifurcation point.	•No gate movements needed. •No communications required. •No decisions needed by operators. •Less silt deposited in main and secondary canals than with other methods because of flow.	•Theoretical equity is not actually achieved because it is virtually impossible to design and install outlets so that they function as predicted over a range of flows. •Even if outlets are precisely designed and installed correctly, hydraulic conditions would change over time (siltation, changing cross sections, changing roughness) yet system can not be adjusted to match changes. • If anything unexpected happens, there is no ability to respond. • Water levels in canals fluctuate greatly with flow-rate changes, causing lining damage and often putting some offtakes above water level.	In warabandi of NW India and Pakistan, lower branches may be ungated and use a rigid rotation. In Bali, Nepal and Fayoum (Egypt) projects are pure flow division.	•If proportional control is needed for rotation farther down the system, main and secondary canals should still be operated with improved control methods to avoid disadvantages listed. •Assumption that simple, nonadjustable proportional control will prevent tampering ignores reality that adjustments must be made over time to fine-tune any field system because of design oversights and changing hydraulic parameters in order to achieve quality.	•Numerous, NW India and Pakistan
2. GATED UPSTREAM CONTROL ▸with flow-rate control	•Typically sluice or radial gates. •Manual operation	Structure should maintain a constant upstream water level. Instead, operators are instructed to use structures as flow-control devices.	None	This is a very common misapplication of upstream control. Operators are asked to do the impossible: with upstream control a constant flow rate cannot be obtained both through the turnout and the check structure.			
▸with water-level control		• First gate, at inlet to canal (and generally an undershot or orifice-type gate) controls flow into canal. • No other check structures control flow; they control upstream water level. • All structures are independently controlled.	• Low initial cost relative to more modern control techniques. • Operation instructions for structures in main canal should be extremely simple: just keep upstream water level constant.	• Flows must be known in advance and be well controlled at inlet and all outlets to minimize tailender problems. • Excellent person-to-person communication needed among canal operators if deliveries have flexibility in order to know how much and when to adjust the turnouts (offtakes) from the main canals. • Tailender problems are pronounced.	• Suitable for arranged deliveries, rotations or proportional control from tertiary canals. • 1-2 day delivery flexibility to offtakes (tertiary canals) is only possible if single canals are shorter than about 50 km, unless large buffer reservoirs are used throughout the system.	• Install buffer (balancing) reservoirs throughout system. • Have large canal sections; use special head-insensitive turnout design to allow pool storage to be varied for operational flexibility. • Install remotely monitored water-level sensors at tailends of canals and at buffer reservoirs.	

(continued on next page)

Annex 1 (continued)

Operational concept	Main canal structure	Function of structure	Advantages	Disadvantages	Applications or design notes	Practice to improve operation	Project and country
▸with structures for manual operation	•Sluice gates (underflow). •Motorized or manual movements			• Large forces required to move the gates. • Often stick in place and become inoperable. • Hourly adjustments needed. • U/S head varies greatly with change in flow, resulting in changing turnout flow rates.		Use side weirs with constant spill (back into the canal) to reduce number of gate movements.	
	• Radial gates (underflow). • Motorized manual movements		• Small forces required to move them if counter-balanced; they do not stick easily.	• Hourly adjustments needed. • Upstream head varies greatly with change in flow.		• Use several smaller parallel gates rather than a few large gates. • Use side weirs with constant spill (back into canal) to reduce number of gate movements.	•Rio Sinaloa and Yaqui, Mexico
	Stoplogs		• Only a few minor changes needed per day. Upstream head varies much less with change in flow than if underflow gate is used.	• Stoplogs may be stolen for firewood. • Stoplogs can become stuck and if too large can be difficult to remove and replace. • Often the walkway above the structure is unsafe for the operator.		• Use stoplogs with maximum dimensions of 2m x 5cm x 10 cm to facilitate handling. • Construct a very stable catwalk for operator convenience and safety.	•Madera ID, California and many western districts, USA
	Long-crested weirs		• Upstream head variations during a day may be almost negligible. • Almost no operator intervention needed. • Extremely simple.	• Do not allow for different controlled water levels during different flow regimes; will silt up unless underflow gates are provided at downstream ends.	Maximum effective design length is about 8-10 times the channel width.	• Install underflow gates at downstream points in each structure to flush silt through the structure. • Adjust the opening of underflow gates if major flow-rate changes occur in canal to minimize the upstream head variation.	•Mocambinho, Brazil •Mae Tang, Thailand •Kemubu, Malaysia •Coello, Colombia
▸with structures for automatic operation	Automatic electrical controls, undershot or overshot gates		• Able to maintain very precise upstream water levels automatically • Target depth can be changed	• Power outages or poorly trained or supplied maintenance personnel will result in failure. • Preventive maintenance is crucial. • Sometimes controllers are too complicated for operators (and maintenance personnel) to understand and adjust. • If control program is not correct, gates will cycle badly, especially if installed in series.	• Can be monitored and controlled remotely in case target depth is changed, or in an emergency. • Make certain that controllers can be adjusted in field and have manual (electrical) over-ride. • Gates must also be movable by hand if electricity fails.	• Use industrial-grade controllers and water-level sensors. • Only use equipment and programs with a proven history of success. • Use side weirs with continuous overflow if gates are undershot. Only move one gate at a time (if several gates are parallel at a site).	•Munda, Malaysia •Friant-Kern Canal, California, USA •Imperial ID, California, USA

Operational concept	Main canal structure	Function of structure	Advantages	Disadvantages	Applications or design notes	Practice to improve operation	Project and country
	Hydraulic constant upstream level gates.		• Very simple. • Almost no maintenance. • For some types, no adjustments are needed after initial installation. • Sturdy and reliable.	• May be greater initial cost than electrically controlled automatic gate. • Water-level control is within a design decrement (does not have the precision of control of electrical controllers). • Target (controlled) water depth cannot be changed on many of these gates.	• Decrement can be reduced if small gates are used in parallel rather than one large gate.	• To reduce price, install one automatic gate in parallel with some manual gates. Manual gates can be adjusted for large flow-rate changes; hydraulic gate can handle daily or hourly fluctuations.	•Sorraia, Portugal •Benil Amir, Morocco •Dudly Ridge WD, California, USA
3. DOWNSTREAM CONTROL WITH TOP LEVEL CANALS	•Gates are always automatic •Either electric or hydraulic	•Maintains a constant water level immediately downstream of gates thereby supplying flow into downstream pool, as needed. • For ordinary operation, flow rates into and through the main canals may not be known. • Flow rates may be checked to see if there are capacity problems.	• Offtakes can be shut off or flows reduced at any time without advance notice. • Very simple, reliable operation of main canal systems. • In effect, this method of operation puts a reservoir at head of each tertiary or distributary canal. • Tailender problems are eliminated.	• Longitudinal slope should generally be less than 0.0003. • Higher construction costs than upstream-controlled canals because of large cross sections needed and level tops. • Operators of offtakes from these canals must generally be very responsible, or they will withdraw more water than canal can supply. If capacity of main canal is sufficient, this is not a problem.	• Demand operation of main and secondary canals is not to be confused with on demand deliveries to individual farmers, chaks, or watercourses. Those deliveries are generally still scheduled or may even be operated on rotation. • The need for human communication to operate main and secondary canals is almost eliminated.	• Turnouts should be located at headends of each pool rather than at tailend of pools. • Emergency siphons or spillways (escapes) must be installed upstream of each structure in case of failure of next upstream check structure, or in case of drainage water entering canal when demand is low.	
	Automatic electrical controls with undershot or overshot gate.		Same as upstream electrical.	Same as upstream electrical.			
	Hydraulic constant downstream level gates		Same as upstream hydraulic gates.	Same as upstream hydraulic gates.			•Sidorejo, Indonesia •Massa, Morocco •Retail Office du Niger, Mali •Tranquility ID, California, USA •Victoria, Australia •Bas Rhone-Languedo, France

(continued on next page)

Annex 1 (continued)

Operational concept	Main canal structure	Function of structure	Advantages	Disadvantages	Applications or design notes	Practice to improve operation	Project and country
4. UPSTREAM & DOWNSTREAM COMBINED CONTROL	• Automatic upstream and downstream control hardware. • A buffer reservoir must exist in main canal at point where upstream control shifts to downstream control.	• First gate, at inlet to upper main canal, is used for flow rate control into system. • All other gates in system only provide water-level control. • Buffer reservoir stores or releases incremental volume differences between anticipated system demand and actual demand. • Inlets to secondary canals can be operated with a very high degree of arranged flexibility as all discrepancies will be absorbed in buffer reservoir.	Less expensive than a complete downstream-control system, yet with about same simplicity and advantages.	• System must be operated on an arranged basis (more restrictive schedules are also possible) because flow into the top of canal is based upon approximate anticipated demands. • Requires a large buffer reservoir in the system (enough for 1-2 days of operational volume discrepancy between orders and actual deliveries).	• Ideal for a canal with an initial steep slope that ends on flatter topography. • Buffer reservoir should be located to side of main canal rather than having full canal flow pass into it (for example, to reduce sedimentation). • Flow-rate changes into canal inlet are based upon daily changes in orders, plus observations of buffer reservoir storage.	• Use modeling to predict wave travel time from inlet to buffer reservoir. • Make 2 to 3 changes in canal inlet flow rate per day (based upon buffer reservoir water level), rather than only once per day.	•Friant-Kern, California, USA •Doukkala Sidi-Bennour, Morocco
5. CENTRALIZED CONTROL ▸with non-responsive scheduling	Often manually upstream controlled gates.	Operators are told by central control how to operate gates for each day's needs, which are generally predicted by some model in central office.	• Central office does not need to listen to field.	• Rarely if ever works as intended, because the control is "open-looped" without any feedback; design and operation assumptions are usually incorrect. In order to even partially work, extensive field calibration of hydraulic parameters must be done.	This is not really a control technology but rather a method of management. It is frequently proposed, however, in recent literature as a means of control.		NE Irrigation, Thailand Upper Pampanga, Aurora-Penaranda, Philippines
▸with arranged delivery	• Electrically controlled, automated gates • Micro-processors at each gate • All gates are electrically moved from a remote centralized control center • Turnouts are generally not automated, nor are they remotely controlled	• All gates respond to commands from a centralized control center. • Gates may maintain water levels or pool volumes. • Central office may use some transfer functions and prediction techniques to send water down canal in anticipation of orders	• Allows fast response throughout system in case of an emergency; all gates can be shut down quickly. • Response time in some systems is theoretically the wave travel time across one pool rather than along whole canal length, as gates can be moved simultaneously.	• These methods generally require 1-2 days' advance notice of any turnout flow-rate change responsive. • Flows are generally input into a simulation program that estimates proper gate settings. Those gate settings are often changed manually from the remote, centralized location. • Requires extremely dedicated, well-trained, and well-funded staff, maintenance program, communications system, and equipment and sensors. • Generally these techniques do not maintain constant water levels in pools.	•Suitable for very large canals and primarily for conveyance. • Especially valuable for areas prone to earthquakes and flooding where quick shutdown of canals is important.	Use same hardware and communications system, but modify the control logic to utilize dynamic regulation (explained below).	• California Aqueduct, USA • Central Arizona Project, USA

Operational concept	Main canal structure	Function of structure	Advantages	Disadvantages	Applications or design notes	Practice to improve operation	Project and country
6. RESPONSIVE SYSTEMS FOR SLOPING CANALS	• Electrically controlled, automated gates • Micro-processors at each gate • Some methods have independently controlled gates; others are moved together. • All these systems need centralized monitoring.	• Responds to computer instructions. • Some maintain water levels; others maintain pool volumes.	• Similar in function to downstream control on level tops. System will automatically provide water to downstream pools as needed, without human intervention and without knowledge of flow rates. • Fast and automatic response to flow-rate increase or decreases at the offtakes. • Minimal human intervention needed for actual operation. • Canal cross sections can be smaller than for level-top canals.	• High risk if personnel, maintenance, initial equipment quality, power backup, communications are not superb. • Require a high degree of initial planning and modeling work and sophisticated maintenance and operation personnel.	Centralized dynamic regulation methods may be compatible with inline hydroelectric installation operations; independent control methods are not.	Large canal cross sections and buffer reservoirs always make control easier even with sophisticated modeling and control.	
▸with local independent controllers	• Radial gates are generally used. • Remote monitoring is highly recommended.	• Controls flow rate into a pool in order to maintain a specified water level at some point in that downstream pool (that is, they operate on demand). • BIVAL maintains level at midpoint; CARDD and EL-FLO maintain level in downstream end of pool.	• Small, relatively inexpensive controllers.	• These control methods have not had wide application. Knowledge is still being gained regarding design rules and requirements for simplicity and techniques for determining proper control algorithm constants.	• Methods do not appear to work well on steep slopes. • Since they are still in development, a full history of past research and applications is advised before use. • They are listed, however, because they appear to be very promising once theory becomes transferable and rules for design and limitations are known.		•Tehema-Colusa, California, USA •Canal du Sahel, Niger
▸with dynamic regulation	• Centralized, computerized control center. • Radial gates are generally used.	• Controls flow into pool in order to maintain desired water level or pool volume. • Movement of any single gate is calculated in conjunction with other gate movements.	• Very fast, responsive operation. • Capable of complex operations, such as integration of pumping stations, reservoirs and hydropower generation along the canals. • Potentially capable of all advantages of independent controllers and centralized arranged systems combined.	Highly sophisticated equipment.	• Several successful systems are in place. • Proven technology.		•Canal de Provence, France •Canal de Haouz, Morocco

(continued on next page)

Annex 1 (continued)

Operational concept	Main canal structure	Function of structure	Advantages	Disadvantages	Applications or design notes	Practice to improve operation	Project and country
7. PRESSURIZED SYSTEM	Closed pipe system	• For main and secondary distribution, pipelines are generally high-pressure pipe. • Similar to municipal system.	• Highest conveyance efficiency. • Minimal maintenance if properly designed and installed, and low silt levels in water. • No spill. • Simple operation unless complex pumping is needed. • Minimal land out of production. • Easy cross-section.	• May require expensive pumping. • Initial investment generally higher than canals. • Pressure regulators are necessary at turnouts because pressures may fluctuate hourly because of flow changes from turnouts.	• Automatic screening needed at entrance to prevent inlet blockage and subsequent pipe damage during refilling. • Adequate pressure relief and air venting designs needed.	• Common problem is to undersize the pipes; systems can be very flexible if pipes are large enough. • Ideally suited for volumetric deliveries. • Flow measurements of turnouts are simple if water is screened at pipe inlet.	•Westlands WD, Belridge WD, Wheeler Ridge WD in California, USA •Nebhana, Tunisia

Distributors of World Bank Publications

ARGENTINA
Carlos Hirsch, SRL
Galeria Guemes
Florida 165, 4th Floor-Ofc. 453/465
1333 Buenos Aires

AUSTRALIA, PAPUA NEW GUINEA, FIJI, SOLOMON ISLANDS, VANUATU, AND WESTERN SAMOA
D.A. Information Services
648 Whitehorse Road
Mitcham 3132
Victoria

AUSTRIA
Gerold and Co.
Graben 31
A-1011 Wien

BANGLADESH
Micro Industries Development
Assistance Society (MIDAS)
House 5, Road 16
Dhanmondi R/Area
Dhaka 1209

Branch offices:
Pine View, 1st Floor
100 Agrabad Commercial Area
Chittagong 4100

BELGIUM
Jean De Lannoy
Av. du Roi 202
1060 Brussels

CANADA
Le Diffuseur
151A Boul. de Mortagne
Boucherville, Québec
J4B 5E6

Renouf Publishing Co.
1294 Algoma Road
Ottawa, Ontario
K1B 3W8

CHILE
Invertec IGT S.A.
Av. Santa Maria 6400
Edificio INTEC, Of. 201
Santiago

CHINA
China Financial & Economic
Publishing House
8, Da Fo Si Dong Jie
Beijing

COLOMBIA
Infoenlace Ltda.
Apartado Aereo 34270
Bogota D.E.

COTE D'IVOIRE
Centre d'Edition et de Diffusion
Africaines (CEDA)
04 B.P. 541
Abidjan 04 Plateau

CYPRUS
Center of Applied Research
Cyprus College
6, Diogenes Street, Engomi
P.O. Box 2006
Nicosia

DENMARK
SamfundsLitteratur
Rosenoerns Allé 11
DK-1970 Frederiksberg C

DOMINICAN REPUBLIC
Editora Taller, C. por A.
Restauración e Isabel la Católica 309
Apartado de Correos 2190 Z-1
Santo Domingo

EGYPT, ARAB REPUBLIC OF
Al Ahram
Al Galaa Street
Cairo

The Middle East Observer
41, Sherif Street
Cairo

FINLAND
Akateeminen Kirjakauppa
P.O. Box 128
SF-00101 Helsinki 10

FRANCE
World Bank Publications
66, avenue d'Iéna
75116 Paris

GERMANY
UNO-Verlag
Poppelsdorfer Allee 55
D-5300 Bonn 1

HONG KONG, MACAO
Asia 2000 Ltd.
46-48 Wyndham Street
Winning Centre
2nd Floor
Central Hong Kong

HUNGARY
Foundation for Market Economy
Dombovari Ut 17-19
H-1117 Budapest

INDIA
Allied Publishers Private Ltd.
751 Mount Road
Madras - 600 002

Branch offices:
15 J.N. Heredia Marg
Ballard Estate
Bombay - 400 038

13/14 Asaf Ali Road
New Delhi - 110 002

17 Chittaranjan Avenue
Calcutta - 700 072

Jayadeva Hostel Building
5th Main Road, Gandhinagar
Bangalore - 560 009

3-5-1129 Kachiguda
Cross Road
Hyderabad - 500 027

Prarthana Flats, 2nd Floor
Near Thakore Baug, Navrangpura
Ahmedabad - 380 009

Patiala House
16-A Ashok Marg
Lucknow - 226 001

Central Bazaar Road
60 Bajaj Nagar
Nagpur 440 010

INDONESIA
Pt. Indira Limited
Jalan Borobudur 20
P.O. Box 181
Jakarta 10320

IRAN
Kowkab Publishers
P.O. Box 19575-511
Tehran

IRELAND
Government Supplies Agency
4-5 Harcourt Road
Dublin 2

ISRAEL
Yozmot Literature Ltd.
P.O. Box 56055
Tel Aviv 61560

ITALY
Licosa Commissionaria Sansoni SPA
Via Duca Di Calabria, 1/1
Casella Postale 552
50125 Firenze

JAPAN
Eastern Book Service
Hongo 3-Chome, Bunkyo-ku 113
Tokyo

KENYA
Africa Book Service (E.A.) Ltd.
Quaran House, Mfangano Street
P.O. Box 45245
Nairobi

KOREA, REPUBLIC OF
Pan Korea Book Corporation
P.O. Box 101, Kwangwhamun
Seoul

Korean Stock Book Centre
P.O. Box 34
Yeoeido
Seoul

MALAYSIA
University of Malaya Cooperative
Bookshop, Limited
P.O. Box 1127, Jalan Pantai Baru
59700 Kuala Lumpur

MEXICO
INFOTEC
Apartado Postal 22-860
14060 Tlalpan, Mexico D.F.

NETHERLANDS
De Lindeboom/InOr-Publikaties
P.O. Box 202
7480 AE Haaksbergen

NEW ZEALAND
EBSCO NZ Ltd.
Private Mail Bag 99914
New Market
Auckland

NIGERIA
University Press Limited
Three Crowns Building Jericho
Private Mail Bag 5095
Ibadan

NORWAY
Narvesen Information Center
Book Department
P.O. Box 6125 Etterstad
N-0602 Oslo 6

PAKISTAN
Mirza Book Agency
65, Shahrah-e-Quaid-e-Azam
P.O. Box No. 729
Lahore 54000

PERU
Editorial Desarrollo SA
Apartado 3824
Lima 1

PHILIPPINES
International Book Center
Suite 1703, Cityland 10
Condominium Tower 1
Ayala Avenue, H.V. dela
Costa Extension
Makati, Metro Manila

POLAND
International Publishing Service
Ul. Piekna 31/37
00-677 Warzawa

For subscription orders:
IPS Journals
Ul. Okrezna 3
02-916 Warszawa

PORTUGAL
Livraria Portugal
Rua Do Carmo 70-74
1200 Lisbon

SAUDI ARABIA, QATAR
Jarir Book Store
P.O. Box 3196
Riyadh 11471

SINGAPORE, TAIWAN, MYANMAR, BRUNEI
Gower Asia Pacific Pte Ltd.
Golden Wheel Building
41, Kallang Pudding, #04-03
Singapore 1334

SOUTH AFRICA, BOTSWANA
For single titles:
Oxford University Press
Southern Africa
P.O. Box 1141
Cape Town 8000

For subscription orders:
International Subscription Service
P.O. Box 41095
Craighall
Johannesburg 2024

SPAIN
Mundi-Prensa Libros, S.A.
Castello 37
28001 Madrid

Librería Internacional AEDOS
Consell de Cent, 391
08009 Barcelona

SRI LANKA AND THE MALDIVES
Lake House Bookshop
P.O. Box 244
100, Sir Chittampalam A.
Gardiner Mawatha
Colombo 2

SWEDEN
For single titles:
Fritzes Fackboksforetaget
Regeringsgatan 12, Box 16356
S-103 27 Stockholm

For subscription orders:
Wennergren-Williams AB
P. O. Box 1305
S-171 25 Solna

SWITZERLAND
For single titles:
Librairie Payot
Case postale 3212
CH 1002 Lausanne

For subscription orders:
Librairie Payot
Service des Abonnements
Case postale 3312
CH 1002 Lausanne

THAILAND
Central Department Store
306 Silom Road
Bangkok

TRINIDAD & TOBAGO, ANTIGUA BARBUDA, BARBADOS, DOMINICA, GRENADA, GUYANA, JAMAICA, MONTSERRAT, ST. KITTS & NEVIS, ST. LUCIA, ST. VINCENT & GRENADINES
Systematics Studies Unit
#9 Watts Street
Curepe
Trinidad, West Indies

UNITED KINGDOM
Microinfo Ltd.
P.O. Box 3
Alton, Hampshire GU34 2PG
England

FRIENDLY–USER CONTROL STRUCTURES

1. **IRAN**
Guilan project. Long–crested weir on Fumen canal.

2. **MEXICO**
Composite cross–regulator.
The radial gates are used for large flow adjustments whereas the side weirs compensate normal flow fluctuations.

3. **DOMINICAN REPUBLIC**
Yaque del Norte project. This Begemann counter–weight gate installed upstream of a drop structure provides nearly constant level and constant flow at farm offtakes.

4. **INDONESIA**
Kedung Ombo Project. Sidorejo Canal.
This 13 km long canal is automatically operated under downstream control.

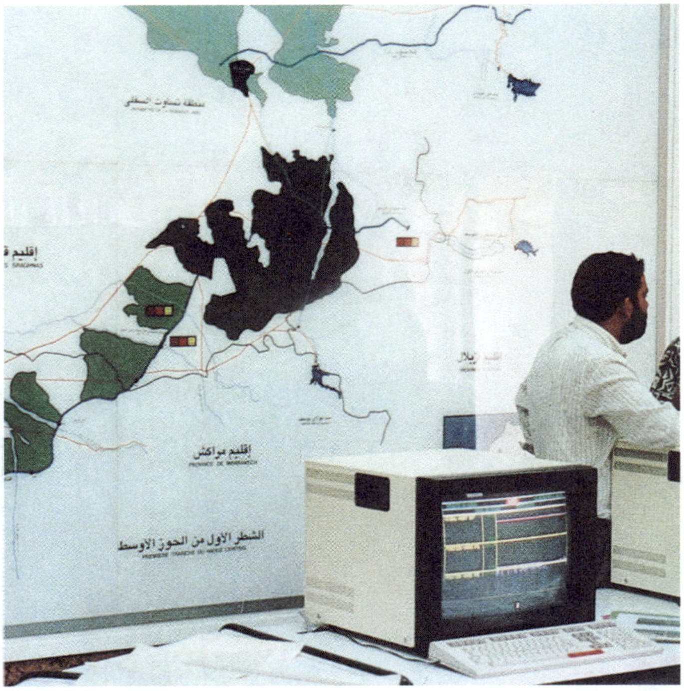

5. Control Center.

6. A composite regulator consisting of a side weir and a remotely–controlled slide gate.

MOROCCO

Haouz Irrigation Project. Canal de Rocade under dynamic regulation.

7. **PERU**

La Joya Irrigation District. Flow divider on the main canal.

SOME STRUCTURES TO AVOID

8. CHO (Constant–Head Orifice) offtakes:
Because of the relative complexity of operation.

9. Rominj gate:
Because of the large fluctuations of the flow rate with minor changes in the water levels of the supply canal.
As shown here, sluice–gated cross–regulators combined with Rominj–gated offtakes is
the worst of all combinations of structures from the hydraulic point of view.

FROM CONVENTIONAL TO INNOVATIVE DESIGN

10. Conventional design

11. Construction of long–crested weir

12. Modular distributor

MEXICO

Tepalcatepec–Cupatizio Project.

The conventional gated structures were modified by replacing one or two slide–gates by long–crested weirs and replacing CHO gates at offtakes by fiberglass modular distributors.